T0315497

Ecological
Parasitology

Ecological Parasitology: Reflections on 50 Years of Research in Aquatic Ecosystems

Gerald W. Esch

Charles M. Allen, Professor of Biology
Department of Biology
Wake Forest University
Winston-Salem, North Carolina, USA

WILEY Blackwell

Dedicated to the memory of my friend, Professor Ralph D. Amen, who died just a few months ago following a friendship of some 50 years. Unfortunately, my "personal editor" was not able to help me this time.

Contents

Preface

From the beginning of my academic career nearly 50 years ago, I have been blessed by strong associations with quality graduate and undergraduate students. Throughout this period, it has been my pleasure to watch them successfully apply their efforts in both the field and the laboratory.

Over the past several years, I have given serious thought to writing about their research experiences and my connection to them. Ward Cooper was a Senior Commissioning Editor at Wiley-Blackwell. I have known him for many years, and he has enthusiastically helped me with other projects along the way. A couple of years ago, I persuaded Ward that stories I wanted to tell about my students and their research were estimable and worthy of incorporating into a book. He agreed and was able to convince the "editorial group" at Wiley-Blackwell to support its publication.

Necessarily, some of what I have written for the present book also involves my professional beginning, making the first couple of chapters seem like the book will be autobiographical, but it isn't. As I will emphasize later, there is no way of separating my career from the beginning of theirs. They are naturally intertwined. I have also included descriptions of contributions made to my career and to those of my students by some of my nonstudent collaborators, mentors, and colleagues. The group includes some really great people, for example, Robert ("Doc") Stabler, Mary Alice Hamilton, J. Teague Self, Jim McDaniel, MacWilson Warren, Jim Hendricks, George Lauff, Desmond Smyth, Clive Kennedy, Whit Gibbons, Darwin Murrell, Al Shostak, Ray Kuhn, Ron Dimock Jr., Dick Seed, and Al Bush. All of these people, plus some others who I will identify in the text, have touched my personal and professional life in one way or another, and I feel that I am in debt to them for contributing to whatever successes my students and I have had along the way.

Foremost, however, thoughts regarding my career always return to my students and what they did in the way of research when they were graduate students with me here at Wake Forest. Without exception, they were bright, self-starters, hard working, innovative, and loyal. During my career, I take pride in knowing that I have had just two students leave my lab without achieving the success of securing their advanced degree.

In 1966, the American Society of Parasitologists held our annual meeting in San Juan, Puerto Rico. I recall it quite well. For one thing, it was the first time I had ever flown. My wife, Ann, and I drove to Baltimore and stayed with our friends, Bill and Sally Cline, the night before we were to head for San Juan aboard an Eastern Airlines Electra turboprop aircraft. My wife was a veteran flyer and felt it would be best to drive to Baltimore and then fly from there rather than taking my first plane trip in a smaller aircraft. Moreover, Bill was a physician and promised to provide me with some sort of remedy that would calm my nerves on flight day. Unfortunately, on arrival at the airport, we learned that a windshield on the plane had cracked on landing in Newark on its way to Baltimore and was about four hours late—by that time, the effects of the drug had worn off. I must admit that it was still a great first flight!

The best part of the San Juan meeting was a dinner on the veranda of the Americana Hotel in which we were staying. Dr. and Mrs. Self, John and Karen Janovy, Jim and Sue McDaniel, Horace and Jayne Bailey, and my wife, Ann, and were there (Figure 0.1) (John, Jim, Horace, and I were all Dr. Self's graduate students). The evening was warm, but not hot. There was a gentle breeze blowing in off the ocean, and the sun was setting in the west. In other words, it was beautiful!

Figure 0.1 A photo taken at the Americana Hotel, San Juan, Puerto Rico, November 1966, 41st annual meeting, the American Society of Parasitologists. From left to right: me, Ann Esch, Horace Bailey, Jayne Bailey, Sue McDaniel, Jim McDaniel, Mrs. Ida Self, Dr. J. Teague Self, and Karen Janovy. Photo courtesy of John Janovy, Jr.

Dr. Self was feeling pretty good (he was a scotch drinker) but was in very good control. We talked about the importance of professional meetings and why we students should attend regularly. I remember Dr. Self nostalgically looking around the table before saying, "I really enjoy coming to these meetings. It gives me the opportunity to bask in the reflected glory of my students." This statement is the perfect description of well-deserved pride in the success a student brings to their mentor, whether at a professional meeting or in the research they accomplish. I have never forgotten this feeling. It is another reason I wanted to write this book.

Acknowledgments

I want to especially thank my many colleagues who, over the years, have encouraged, guided, and advised me in every way possible. Their gift to me taught me to give as much to my students. As mentoring models, I could not have had a better bunch.

I also want to thank my students for their cooperation in helping me to write this tome, in correcting a number of errors, and in reminding me of things that I had omitted. Their company and friendship have been outstanding ever since they walked into my lab and participated with me whether it was at Colorado College, the University of Oklahoma, Wake Forest University, Kellogg Biological Station, Imperial College in London, or the Savannah River Ecology Laboratory. My final graduate student, Kyle Luth, was particularly helpful. He read the entire book and offered a bushel full of suggestions, most of which I gratefully accepted. Just before submitting the final version to Wiley-Blackwell, I asked a long-time friend, Dr. Richard Seed, to read the book and offer any suggestions he might have for improving the content. He was very helpful.

The people at Wiley-Blackwell, especially Ward Cooper and Kelvin Mathews, helped put it together. My project manager, Jayavel Radhakrishnan, provided very good technical assistance.

Finally, I thank my children, Craig, Lisa, and Charlie, who willingly accompanied me almost all the way on this long journey. Most especially, I am grateful to my wonderful wife, Ann, who has been so fantastic throughout our lives. She is one of my best, and toughest, critics. Though not a scientist, she read the entire final draft and offered a significant number of ideas for improving the content and the way I presented it.

The Players

There never was in the world two opinions alike, no more than two hairs, or two grains; the most universal quality is diversity.

Of the Resemblance of Children to Their Fathers,
Michel de Montaigne (1533–1592)

As I approached the end of the present manuscript, I realized that the book was not just a succession of individual student stories; in many cases, they were clearly linked to one another. It was like my students and I had traveled together on a journey. As each of them came on the scene, they stayed with me for a while and then departed to begin their own travels. A few have retired, but most are still making their way. Many of them have become connected to their own students for part of their trips in the same way they joined with me and I with Dr. Self, my graduate school mentor. I hope I have been as successful as a role model as he and Dr. Stabler, my undergraduate guide, were for me. I must emphasize again that the book is not an autobiography—it is a collection of intertwined biographies, with mine included.

Early, while writing their stories, I began preparing microbiographies for each of student who had taken an ecological route—there were several who did not follow in this research direction and are not included. The original idea was to include their narratives at the end of the book, but I decided to move this information to the beginning. This way, you the reader would have at least an idea about each of them as I begin to tell their stories.

During my 50 years as a faculty member here at Wake Forest University, I have had the privilege of mentoring 24 master of science (MS) and 19 PhD students. Of the MS students, five have continued on for their PhD degrees with me. Unfortunately, I lost track of seven students after their graduation. However, I know that at least one in this group became a dentist, another became a physician, two are housewives, one earned a PhD elsewhere, and two others work in the corporate world.

I consider all of my students as close friends, and I am truly pleased to see a great many of them every year at the annual meeting of the American Society of Parasitologists. Ray Kuhn, my longtime friend and colleague, has graduated about the same number of students as I, and a great many of his former students attend the annual meeting as well. I think the largest combined number was 17, at the ASP meeting in Colorado Springs in 2010. One advantage of having such a large group is the extent of "networking" that is accomplished by the students after they leave here—it really means a lot!

Microbiographies

John Trainer

John was my first graduate student—he was in fact waiting for me when I arrived at Wake Forest in the fall of 1965. So, we began our journey together. Although his research did not involve parasites in aquatic ecosystems, collecting for his thesis was accomplished in the field. After acquiring his MS degree, he went on to the University of Oklahoma where he earned his PhD with Dr. J. Teague Self (my mentor in graduate school). John's first academic position was at Jacksonville University (Jacksonville, Florida) where he moved rapidly through the academic ranks, ending up as vice president for academic affairs and dean of faculty after just 9 years in residence. His career as an administrator then took him to Lenoir–Rhyne College (Hickory, North Carolina) as president. Eventually, he was to become president and head of The Bolles School in Jacksonville, one of the finest K–12 private schools on the east coast of the United States. After his retirement at The Bolles School, he returned to Jacksonville University where he presently serves as the senior campaign officer. He recently sent me a yearbook that included photographs, warm reminisces and recollections, etc., dealing with his retirement gala at Bolles—what an exciting experience that must have been for him and his wonderful wife, Alice!

Robert (Bob) Morcock

Bob was one of my first PhD students. He began his doctoral work in 1969 right after the PhD program was established in our Department of Biology at Wake Forest. His dissertation research was focused on the intermediary carbohydrate metabolism of larval *Taenia crassiceps* (about as "nonecological" as one could get). Although I had not yet made the switch to ecological parasitology, I was able to provide him some help since my own dissertation work was aimed at comparative intermediary carbohydrate metabolism in larvae and adults of *Taenia multiceps*. He is nonetheless included in my ecology student group because it was Bob who inadvertently started our work on the population biology of *Crepidostomum cooperi* in Gull Lake during its eutrophication/recovery process. After completing his degree, he obtained a teaching job at Hood College (Frederick, Maryland). He stayed there for a relatively short time before going to work at the Environmental Protection Agency (EPA) in Washington, DC, where he spent the rest of his career, retiring just recently.

John Aho

John obtained his BS and MS degrees here at Wake Forest. I remember recruiting John into my lab when he was a sophomore. I needed someone (preferably a student—because they came relatively cheap) to care for my mouse colony. While teaching a course in cell physiology, I walked into my teaching lab one day and asked if there was someone who would like to make a few extra dollars by caring for my animals. John's hand shot up immediately. He stayed with me for several years while completing his undergraduate work and his first few years as a graduate student. On completion of his MS degree, he secured a Fulbright scholarship and used it to obtain his PhD degree at the University of Exeter (England) under the tutelage of Dr. Clive Kennedy. I was even invited to be his external examiner for his dissertation. After completing his PhD, John then spent time with Dr. John Holmes at the University of Alberta (Edmonton, Alberta, Canada) as a postdoc and back here at Wake Forest as a visiting professor before heading for Auburn University (Montgomery, Alabama).

Joseph (Joe) Bourque

I first met Joe when I was teaching field parasitology at the W.K. Kellogg Biological Station (Hickory Corners, Michigan) of Michigan State University. Joe came from the University of Illinois (Champaign, Illinois) to Wake Forest to first do an MS degree and then his PhD. He was one of my first two "genuine" parasite ecology students. His research was conducted at the Savannah River Ecology Laboratory (Aiken, South Carolina), where he studied the effects of thermal effluent stress on the population and community biology of helminth parasites in yellow-bellied turtles. After successfully completing his dissertation research and his PhD degree, he joined the Peace Corps and served in southeastern Asia. When he returned home from the Peace Corps (with a wife), he first worked as a medical technician at Northwestern University (Chicago, Illinois) and then was finally was admitted into the University of Illinois Medical School (Urbana, Champaign, Illinois) where he pursued an MD degree. He spent his career as a thoracic surgeon in California.

James (Jim) Coggins

Jim Coggins obtained his BA and MA degrees at East Carolina University (Greenville, North Carolina), where he worked with Dr. Jim McDaniel. The latter, Jim, and I had been graduate students together at the University of Oklahoma (OU) (Norman, Oklahoma). He was the first graduate student I met when I arrived on the campus at the OU in 1958, and we became lifelong friends. Very regrettably, he died way too young, and I still miss him in so many ways. Jim Coggins came to Wake Forest to work on his PhD. His research was focused on the plerocercoid stage of *Proteocephalus ambloplitis*, the so-called bass tapeworm, with primary interest on the fine structure and histochemistry of the apical end organ of this unusual tapeworm. During one summer of his stay at Wake Forest, Jim accompanied me to KBS where he served as my teaching assistant. He and I did some collaborative work using what we called "tethers," which in reality were wide-mesh baskets made of heavy twine in which we could place uninfected bluegills collected from a nearby pond. We planted the "tethers" in selected locations of the Gull Lake littoral zone so that we could learn something about the rates of helminth recruitment and turnover under controlled conditions. It was an interesting undertaking. When Jim finished his dissertation work, he headed off to the University of Notre Dame (Notre Dame, Indiana) and a postdoc with Dr. Paul Weinstein. He continued a successful career at the University of Wisconsin–Milwaukee (Milwaukee, Wisconsin), from which he is now retired.

Herman Eure

I write at length about Herman in the text of Chapter 7, so I will not repeat everything here. Herman came to Wake Forest with his undergraduate degree from the University of Maryland–Eastern Shore (Princess Anne, Maryland) on one hand and a prestigious Ford Foundation Fellowship on the other. He decided to skip the master's degree and head straight for his PhD. His research was conducted on the population biology of the bass tapeworm *P. ambloplitis* and the acanthocephalan *Neoechinorhynchus cylindratus* in Par Pond on the Savannah River Plant (Aiken, South Carolina). The year he finished his PhD, I left Wake Forest to take a leave of absence at the Savannah River Ecology Laboratory (SREL), and Herman was appointed as my "temporary" replacement. On my return the next year, Herman was hired as a tenure-track faculty member in our Department of Biology. As our graduate school's first African-American, in addition to his faculty responsibilities, he was employed in a variety of administrative positions before becoming our Department of Biology's chairman, where he did a superb job. He subsequently married Kelli Sapp and recently retired as a faculty member in the college. He is, however, still active in the American Society of Parasitologists, for which he and Kelli serve as co-program officers.

Terry Hazen

Terry obtained his BS and MS degrees in interdepartmental biology from Michigan State University (East Lansing, Michigan). I first met Terry when he came to the Kellogg Biological Station (KBS) to take my field parasitology course in the summer of 1973. He accompanied me to the SREL as my technician while I was on leave from Wake Forest during the 1974–1975 academic year. When I returned to Wake Forest in the fall of 1975, he came with me to pursue his PhD. His dissertation research focused on the ecology of *Aeromonas hydrophila*, the causative agent for red sore disease in fish. While he did some parasitological research early on, Terry became an internationally renowned microbial ecologist, with strong interests in bioremediation and bioenergy technologies. Throughout his career, he has held several academic/administrative positions of responsibility at the University of Puerto Rico, the Savannah River Plant, and the E.O. Lawrence Berkeley National Laboratory in California (Berkeley, California). He is presently the University of Tennessee/Oak Ridge National Laboratory Governor's chair (Oak Ridge, Tennessee) and professor in their Department of Microbiology and the Department of Earth and Planetary Sciences (Knoxville, Tennessee).

Joseph (Joe) Camp

Joe received BS and MS degrees at Illinois State University (Normal, Illinois), where he worked under the guidance of a longtime friend, Dr. Harry Huizinga. He came to Wake Forest in 1977 to pursue a PhD degree in my lab. His research involved a study on the population biology of *Tylodelphys* (*Diplostomulum*) *scheuringi* in mosquito-fish (*Gambusia affinis*) in Par Pond, on the Savannah River Plant site. On receipt of his degree, he took a position at Purdue University North Central (Westville, Indiana). He subsequently transitioned to Purdue's main campus (West Lafayette, Indiana), where serves as a professor of veterinary medicine and secretary of faculties.

A very coincidental occurrence took place when I traveled to Lawrence, Kansas, to visit with several people with whom I would work at Allen Press (AP) after being appointed successor to Brent Nickol as editor of the *Journal of Parasitology* in 1993. We were at lunch and a very nice lady was sitting across the table. I asked her what she did at AP, and she replied that she was a copy editor for our journal. I inquired if this was her only job and she replied, "Oh, no. I am also the Dean of the College of Liberal Arts and Sciences at the University of Kansas" in Lawrence; I almost choked on the bite of salad I was trying to swallow. I asked why in the world she would hold these two sorts of jobs. She explained that she enjoyed working in the evening while sitting in a comfortable easy chair using a writing board to edit while listening to beautiful music—it "was relaxing," she remarked. I later learned that she (Sally Mason) had become provost at Purdue University. It was she who invited Joe Camp to come down to the main campus to become secretary of the Faculty Senate, as well as professor in their School of Veterinary Medicine where he also teaches parasitology. Sally Mason is presently the president at the University of Iowa!

Willard (Bill) Granath Jr.

Bill completed his undergraduate degree at Delaware Valley College of Science and Agriculture (Doylestown, Pennsylvania) and then his MS degree at Illinois State University (Normal, Illinois) where he worked with Harry Huizinga. Joe Camp and Bill had become friends at Illinois State and Joe strongly recommended that Bill come to Wake Forest for his PhD degree. He applied and was accepted—a great move on our part! His doctoral research was to focus on the population ecology of the Asian tapeworm *Bothriocephalus acheilognathi* in mosquitofish in a nearby cooling reservoir operated by Duke Energy. I am especially glad that Bill came here, because I seriously doubt that any research on *B. acheilognathi* would have ever been undertaken (in fact, I do not believe that we would have known about the presence of *B. acheilognathi* in the reservoir because I still cannot see how this tapeworm can fit into the gut of such small fish!). Moreover, the fish fauna had been significantly reduced via selenium pollution, and the only piscine species left in the main body of the lake were a few carp, channel catfish, and mosquitofish; while still largely devoid of fish, fathead minnows and red shiners were introduced into the lake. Those of us who worked in Belews figure that a frustrated bass fisherman who could not catch anything in the main body of the lake, let alone get a bite, "donated" their baitfish to the lake after a frustrating "no-catch" day. On completion of his PhD degree, Bill did postdoctoral work with Dr. Tim Yoshino at the University of Oklahoma (Norman, Oklahoma) before he obtained a tenure-track position at the University of Montana (Missoula, Montana). He also switched his research focus to the cnidarian *Myxobolus cerebralis* and has become a leading world expert on whirling disease in salmonid fish, definitive (final) hosts for the parasite.

Amy Crews

Amy started as an undergraduate at Emory University (Atlanta, Georgia) before transferring to Wake Forest where she completed her BS and MS degrees. She was the first to work in Charlie's Pond and set a very high standard for the students who followed her into our small impoundment. She began her PhD studies with Dr. Tim Yoshino at the University of Oklahoma and finished at the University of Wisconsin School of Veterinary Medicine (Madison, Wisconsin) after Tim moved from Oklahoma. She then did postdoctoral work with Bill Collins at the Centers for Disease Control (CDC) in Atlanta. Amy accepted a teaching position at the University of Northern Alabama (Florence, Alabama) where she presently works as director of their Health Professions Committee.

Timothy (Tim) Goater

Tim was the first of three Canadian graduate students to come down to the relatively warm south of North America. His undergraduate mentor was Dr. Al Bush at Brandon University (Brandon, Manitoba, Canada). Without question, Tim could not have had a more interesting and knowledgeable person to introduce him to the ecological aspects of parasitology than Al Bush, who, regrettably, died not very long ago at a very young age. For his master's research at Wake Forest, Tim chose to examine the community ecology of salamanders in streams of the Smoky Mountains of southwestern North Carolina and Tennessee. He stayed on to do his PhD at Wake Forest and produced a wonderful dissertation dealing with the trematode *Halipegus*

(=*Hal.*) *occidualis* in Charlie's Pond. It was Tim who also discovered the colonization of *Physa gyrina* and *Halipegus eccentricus* into the pond. On completion of his dissertation, he became a faculty member of what is now known as Vancouver Island University (Nanaimo, British Columbia, Canada).

In 2001, Al Bush, Jackie Fernandez, Dick Seed, and I coauthored a book titled *Parasitism: The Diversity and Ecology of Animal Parasites*. A few years ago, Cambridge University Press, which published the first edition of the book, approached me with an invitation to write a second edition. I told them that I would, but I needed to select a couple of new coauthors to assist since Al had died and Dick and Jackie had retired. Cambridge gave me the "go-ahead," and I asked Tim and Cam Goater (brothers) if they would be interested (I recalled another brother duo, Elmer and Glenn Noble, who had written another parasitology textbook several years ago and how successful it was). Tim and Cam accepted the invitation. They assumed the lead in rewriting the book and, by far, did most of the work in getting the revision completed. While I was writing this section of the present book, a copy of the 2nd edition arrived (Goater et al., 2014). It is an exceptionally fine piece of work (if I do say so myself!).

David (Dave) Marcogliese

Dave was the second "Canuck" to come south. He overlapped with Tim, and together, they were exciting guys to have around—lots and lots of enthusiasm and energy from the pair. Dave did his undergraduate work at Concordia University (Montreal, Quebec, Canada) and then moved to Dalhousie University (Halifax, Nova Scotia, Canada), where he completed his MS He worked for the Department of Fisheries and Oceans for three more years before coming to Wake Forest to begin his PhD studies. While he did some work on *B. acheilognathi*, most of his dissertation research dealt with the biology of copepods, which serve as first intermediate hosts for the parasite. Dave has since gone on with a fantastic career in Environment Canada (something like the Department of Interior in the United States). His research over the years brought the recognition he deserved from the Canadian Society of Zoologists with his receipt of their Robert Arnold Wardle Award and the Henry Baldwin Ward Medal from the American Society of Parasitologists.

Jacqueline (Jackie) Fernandez

Jackie came to Wake Forest from Chile. She had applied to work with my good friend, the late John Crites. However, John was in the process of retiring and recommended that she come to Wake Forest and do her dissertation research with me, which she did. When she arrived, she had more than 20 publications and both undergraduate and MS degrees from the University of Concepcion (Concepcion, Chile). She fit into my laboratory beautifully. Jackie is not only a charming and intelligent young woman, but also she is highly energetic and a very hard worker. Her research mostly involved the infracommunity and component community dynamics of trematodes in *Helisoma (Hel.) anceps* in Charlie's Pond. While here at Wake Forest, she met a young, former marine who was working toward (successfully) his PhD degree in the Department of Chemistry. They were married and had two boys who are now in the process of completing their college educations. Jackie is presently teaching high school biology in Virginia and was recently recognized as teacher of the year by the National Association of Biology Teachers, a really great honor!

A. Dennis Lemly

Dennis started out with a strong interest in wildlife biology and a BS degree from Western Carolina University (Cullowhee, North Carolina). His initial career goal was to become a wildlife and fisheries officer. Toward that end, he graduated with an applied science degree from the Haywood Community College (Clyde, North Carolina) in 1975. Then, he considered teaching as a profession and graduated from Wake Forest University with an MAEd. degree. Dennis was a superior student and was lucky that he already had his dissertation research set up when he began his PhD program in my lab. In the second year of his master's degree work, he took my parasitology course, and his focus changed again, this time to parasitology and ecology. For his PhD research, he centered on the population biology of *Uvulifer ambloplitis*, the causative agent of "black spot" in fish. To my way of thinking, his doctoral research represents a classic contribution to understanding the concept of overdispersion by parasite infrapopulations. He also began a lifelong career concentrating on the effect of selenium pollution in aquatic ecosystems and has become a "world-class" expert in this area of research. He is presently employed by the US Fish and Wildlife Service and has an appointment as research associate professor at Wake Forest.

Michael (Mike) Riggs

Mike Riggs obtained his undergraduate training at the University of New Hampshire (Durham, New Hampshire) and then his master's degree at Iowa State University (Ames, Iowa) where he worked with Dr. Martin Ulmer. He came to work toward his PhD in my lab at Wake Forest in 1979. His very strong research was conducted on the Asian tapeworm *B. acheilognathi* in several minnow species at Belews Lake. About midway through his PhD degree program, Mike decided to also pursue a master of public health degree with support from a US Public Health Service training grant in biostatistics at the School of Public Health, the University of North Carolina–Chapel Hill (Chapel Hill, North Carolina). He returned to Wake Forest where he completed his dissertation in 1986. He worked as a statistician with the Division of Fish and Wildlife, Minnesota Department of Natural Resources (Minneapolis, Minnesota) and returned as a senior environmental statistician for Research Triangle Institute (RTI) in the Research Triangle Park, North Carolina. Mike is now living back in New England.

Eric Wetzel

Eric is a Pennsylvania native who did his undergraduate degree at Millersville University (Millersville, Pennsylvania) before coming to Wake Forest. He began as a master's student of Dr. Peter Weigl in our Department of Biology. Pete was a long-time worker on flying squirrels in the nearby Appalachian Mountains, and these were the "critters" to which Eric first became attached. Eric's focus was on a species of *Strongyloides* that infects the squirrels, and this is what led him into parasitology and my lab after completing his MS degree. His PhD research took him into Charlie's Pond and an emphasis on *Hal. eccentricus*. After completing his degree, he stayed at Wake Forest for another year and a half as a visiting instructor before securing a teaching position at Wabash College (Crawfordsville, Indiana). It was a perfect fit for Eric since Wabash is one of the finest liberal arts colleges east of the Mississippi River, and this is exactly what he wanted in terms of a teaching career. Recently, Eric has entered into another area of parasitology, one that has led him into Peru in South America where he is focusing on infectious parasitic disease research among the poorest of people in that part of the Western Hemisphere. He is presently the Norman E. Treves professor of Biology at Wabash.

Derek Zelmer

Derek is the third Canadian to come south. His BS and MS degrees were both obtained at the University of Calgary (Calgary, Alberta, Canada), where he studied with the late Dr. H.P. Arai. His PhD degree was broadly focused on several aspects of the life cycle and population biology of *Hal. occidualis*. In collaboration with Eric Wetzel (mentioned earlier), Derek was able to establish that the life cycles of *Hal. occidualis* could be centered at four different 'hotspots' in the pond, all of which had similar microhabitat, or landscape, characteristics. He also clearly determined that odonate naiads are necessary paratenic hosts for the parasite and, finally, that transmission of *Hal. occidualis* in Charlie's Pond is governed by prevalence and not intensity of adult parasites. When Derek finished his PhD research, he was able to secure a teaching position at Emporia State University (Emporia, Kansas), an opportunity that both he and I thought was right for him. Unfortunately, we were both wrong and his life in Emporia was miserable. After 6 years, however, he found a much, much better position at the University of South Carolina–Aiken, and he is quite at home in his recently found surroundings.

Kym Jacobson

Kym came to Wake Forest with a BS degree from the University of Nevada–Reno. She completed her master's degree with me and then switched into immunology and obtained her PhD with Ray Kuhn here at Wake Forest. Her master's degree involved the community biology of enteric helminths in the yellow-bellied slider turtles down at the SREL. She did not, however, enjoy performing turtle necropsies, which led her into Ray's lab. After her PhD, the next stop was at the Mayo Clinic (Rochester, Minnesota) and a postdoc in neuroimmunology. Kym then spent time as a postdoc at the Seattle Biomedical Research Institute of the University of Washington (Seattle, Washington). Despite her success in immunology, I always had a "gut feeling" that the ecological aspects of parasitology were her first love, and I was correct. Accordingly, she accepted a postdoc, followed by a full-time staff position, as a fish ecologist/parasitologist at the National Oceanic and Atmospheric Administration (NOAA) Laboratory in Newport, Oregon. I am so proud that she has created a solid reputation for herself in marine ecology and parasitology.

Julie Williams

Julie Williams completed her undergraduate degree with Dr. Brent Nickol at the University of Nebraska–Lincoln (Lincoln, Nebraska) where she developed an interest in parasitology. She came to Wake Forest and successfully pursued her MS degree in my lab. During her last year here, she explained that she wanted to become a physician, and that is what she did! She attended the Bowman Gray School of Medicine (now the Wake Forest University School of Medicine) and is now successfully employed there in their Department of Geriatrics/Gerontology.

Scott Snyder

Scott was another student from the University of Nebraska–Lincoln, where he was coaxed/persuaded into parasitology by my old friend, Dr. John Janovy Jr. Scott was successful in his MS degree research here at Wake Forest. His research was focused on *Hal. eccentricus* in *P. gyrina* in Charlie's Pond, work that provided a really excellent comparison with its sympatric congener in our pond. He returned to Nebraska and John where he performed some really excellent research on frog trematode biology while working toward a PhD degree. This was followed by stints at the University of Wisconsin–Oshkosh (Oshkosh, Wisconsin) and as a program director in systematics at the National Science Foundation. He is presently a faculty member and administrator at the University of Nebraska–Omaha (Omaha, Nebraska).

Anna Schotthoefer

Anna received her BS degree in biology at Cornell University (Ithaca, New York) in 1993 and then spent a couple of years at the Virginia Institute of Marine Science (Gloucester, Virginia) with Dr. Eugene Burreson. She came to Wake Forest in 1996 and completed her master's research in 1998. Her work was focused on the community structure of digenetic trematodes in the pulmonate snails *Hel. anceps* and *P. gyrina*. She was one of the key players in developing ideas regarding differential transmission foci for parasite species in Charlie's Pond. She completed her PhD with Dr. Rebecca Cole and is presently a housewife in Aurora, Colorado.

Kelli Sapp

Kelli spent a summer at Wake Forest as an NSF REU fellow before acquiring her undergraduate degree at Methodist College (Fayetteville, North Carolina). She successfully completed her MS degree in my lab before heading off to the University of New Mexico (Albuquerque, New Mexico) and pursuing her PhD with Dr. Sam Loker. As is noted in Chapter 7, she is married to another of my former students, Dr. Herman Eure. Presently, Kelli is a very successful faculty member at High Point University (High Point, North Carolina). She and Herman are currently coprogram officers for the American Society of Parasitologists.

Brian Keas

Brian completed his undergraduate degree at Hope College (Holland, Michigan) where he came under the influence of Dr. Harvey Blankespoor, another old friend of mine. He completed his MS degree here at Wake Forest with a very complicated effort to understand the growth and reproductive characteristics of *Hel. anceps* with and without infection by *Hal. occidualis*. He then moved on to East Lansing, Michigan, where he completed his PhD at Michigan State University. He is presently a faculty member at Ohio Northern University (Ada, Ohio).

Giselle Broacha-Bruzinski

Giselle obtained her undergraduate degree at Belmont Abbey College (Belmont, North Carolina). She successfully worked for her MS degree by examining trematode population biology in mosquitofish in Charlie's Pond. She was married just after securing her master's degree and is now a housewife in Pennsylvania.

Joel Fellis

Joel began his undergraduate work at Lewis and Clark College (Portland, Oregon) and then completed it at the University of New Mexico and came immediately to Wake Forest. His research for his MS degree was focused on the community structure and seasonal dynamics of helminth parasites in green and bluegill sunfish in Charlie's Pond. Interestingly, the suites of parasites in the two fish hosts were quite similar, but when their abundance patterns were examined, he found them to be distinctly different. I was quite pleased that he stayed in my lab to do his dissertation research, which involved the effects of the life cycle variation on community assembly and population genetic structure of macroparasites in bluegill sunfish from a series of ponds situated at varying distances from each other in Forsyth

County, North Carolina. Joel decided to pursue a nonacademic career after finishing his PhD degree and has become a very successful businessman in California.

Michael (Mike) Barger

Mike is a "flatland" guy who eventually ended up in the Appalachian Mountains working on his PhD His undergraduate work and MS degrees were obtained at the University of Nebraska–Lincoln, where Drs. John Janovy Jr. and Brent Nickol first mentored him. Lotic habitats include rivers and stream, while lentic habitats are comprised of standing water such as ponds and lakes. Most of the aquatic work involving parasites has been accomplished in lentic settings, so Mike decided that he would become involved with stream systems up in the Appalachian Mountains in the headwaters of the Yadkin River. A couple of unexpected opportunities preceded his stream ecology research, however, when he first encountered *Allopodocotyle chiliticorum* n. sp., which he described. To compound his early venture into the streams, he then found *Plagioporus sinitsini*, a very unusual trematode parasite with a 1-, 2-, or 3-host life cycle, all at once! He eventually moved into landscape community ecology in fish from the headwater streams of the Yadkin River, a most productive experience. On completion of his dissertation research, he found a "perfect" academic position at Peru State College (Peru, Nebraska) back home in his flatland surroundings of southeastern Nebraska.

Lauren Camp

Lauren is the daughter of Joe Camp, a former PhD student of mine. She obtained her undergraduate degree at the University of Chicago (Chicago, Illinois) and came to Wake Forest to acquire her MS degree in my lab. Lauren was the first of my graduate students to do any research with the rhabditid nematode *Daubaylia potomaca* in Mallard Lake, a small pond located on a golf course not far from Winston-Salem. Her work was very thorough and paved the way for Mike Zimmermann and Kyle Luth to subsequently focus on the parasite. On completion of her master's degree, she moved to the University of California–Davis (Davis, California) where she has worked toward her PhD under the direction of Dr. Steve Nadler.

Nicholas (Nick) Negovetich

Nick came to Wake Forest from Wabash College (Crawfordsville, Indiana) and Eric Wetzel (mentioned earlier) in 2001. His MS and PhD degrees both involved population and community ecology of digenetic trematodes in Charlie's Pond. Using mark–release–recapture methods for *Hel. anceps*, he was successful in mathematically modeling the snail population biology using matrix models. After receiving his PhD in 2007, he did a postdoctoral stint over a 4-year period at St. Jude Children's Research Hospital in Memphis, Tennessee, where he worked with Dr. Robert Webster, the noted virologist/epidemiologist. Nick recently accepted an academic position at Angelo State University (San Angelo, Texas).

Kyle Luth and Michael (Mike) Zimmermann

When I began writing the microbiographies of my graduate students, I decided to put Kyle and Mike together. Mike is finished now and presently teaches at Shenandoah University (Winchester, Virginia). Kyle will be my last graduate student; he is analyzing his data and will be done in another year or so. Together, they arrived here in the fall of 2008 and began their master's work on a rhabditid nematode, *D. potomaca*, although Kyle had to pull out of his initial focus because his research was not producing what we expected. They both finished their MS degrees in the spring of 2010.

Kyle Luth's first mentor was none other than Eric Wetzel who did his PhD degree with me here at Wake Forest (see Chapter 7) and then secured a position at Wabash College. Both of these guys and I were similar to a lot of youngsters who begin our undergraduate work thinking about a premed program and medical school. At some point during their undergraduate career, however, they came under the influence of a professor, or a fascinating aspect of biology, which then stimulated a teaching and research career.

Mike graduated from Muskingum College (New Concord, Ohio) in 2008. In addition to playing baseball (he was a pitcher) and football (American style—he was a punter), he became interested in parasitology as an undergraduate student while doing an honors project on fish parasites in several abandoned "sandpits" east of New Concord. He had read some fish parasitology papers coauthored by my graduate students and me while working on his undergraduate degree at Muskingum, and his interest in these parasites was stimulated.

I wanted to end the biographic sketches on a positive note, and I think there is nothing better than to briefly describe the content of their PhD dissertations. When they were completing their master's degrees, I approached each of them with a different research proposal, which captured their attention and both decided to stay, representing the last two of my graduate students. Since they represent the end of my career, I thought it would be of interest to describe their work together because I think it is among the best my students would have accomplished.

Kyle's focus was to be on the geographic differences in biology of autogenic *P. ambloplitis*, the so-called bass tapeworm. The basis for his study is related to the fact that adults of the cestode occur in the summer in the north (Esch et al., 1975) and in the winter in the south (Eure, 1974). The original object was to collect *P. ambloplitis* in the north and south and then perform PCR and sequencing to see if populations are genetically different geographically. In other words, are we dealing with cryptic species, or is it some sort of behavioral, physiological, or ecological phenomenon that can account for the life cycle pattern in the north and south?

Mike's study was to concentrate on the genetic differences of the allogenic trematode *Echinostoma revolutum* in the three flyways used by Canada geese east of the Rocky Mountains. How much genetic mixing and variability occur in these parasites? In Mike's case, this would not require shooting of the geese in the three flyways to collect adults of *E. revolutum*. It was sufficient to collect snails and use echinostome metacercariae for the conduct of PCR, sequencing, and molecular comparisons. In both cases, this is the way their work started.

However, as things progressed early during their research, Kyle developed the idea that it might be a good thing if he could work from east to west as well as north and south. So, he approached me with the proposal and I agreed. Mike quickly joined in because he had already planned on working east to west if he was going to sample in the three flyways. So, why didn't he join with the field efforts proposed by Kyle? After all, four hands are considerably more effective than two! After a huge collecting effort in the summer of 2012, they decided to expand their collecting even further in 2013.

Altogether, they were in the field for 70 days over two summers. They covered greater than 16,000 miles and sampled 235 sites in 231 ponds and lakes in 29 states east of the Rocky Mountains all the way to the Atlantic Coast and south of the Canadian border down to the Gulf Coast. In doing so, Kyle collected 2361 centrarchid fish, and Mike examined greater than 9000 snails, representing five species. I should emphasize that Kyle has also performed full necropsies on all of these fish so that he not only acquired plerocercoids of *P. ambloplitis* but enteric adults and parenteric larval helminths as well. Mike also collected data for cercariae shedding. In other words, the two of them assembled two of the largest, if not the largest, sets of field data of which I am aware. Mike successfully defended his dissertation research in April of 2014. Kyle decided that he wanted to stay on an extra year so that he could complete his genetics work and thoroughly conduct the data analysis for his dissertation.

Closing comments

I was sitting with Mike Zimmermann one afternoon in the spring of 2014 talking about his research—I had just spoken with Kyle 30 min earlier about the same thing. Mike reminded me that in order to successfully accomplish something like the two guys had, they must have a good work ethic, and they both did. I began thinking about this a little more as the afternoon passed.

I agree that a strong work ethic is essential, but there is more. There was also a need for financial backing, and I had no grant money to support their work. Instead, they used money from what is called the Grady Britt Fund, actually named for the man I replaced when I came to Wake Forest. I do not know the circumstances of his departure only that he went on to a small college in the Deep South where he finished his career. He had no family, and when he died, he left his estate to the Department of Biology with instructions that income generated was to be used by graduate students in parasitology. With support from the Grady Britt Fund, Kyle and Mike had dollars to subsidize their field trips into the "boonies" and then to handle their PCR and sequencing work.

I also believe these two students, plus all the others I have directed, had to have an "intangible asset" that people must possess who pursue either or both of the two highest graduate degrees. The intangible is something deep inside a person. It is difficult to define, but I know it is there. Personally, I can sense it every time I see one of my students achieve success. I know it is there because it is a feeling I have when one of my students secures a job, or publishes a paper, or is admitted into another graduate program, or enters medical school. There is a quotation I used once before during my career (Esch, 1999), and I believe that it is a reflection of my personal intangible. It comes from Carolus Linnaeus (1707–1778). He is quoted, "A professor can never better distinguish himself in his work than by encouraging a clever pupil, for the true discoverers are among them, as comets are among the stars." While my personal intangible is difficult to define, I am certain that Linnaeus' declaration adequately expresses it.

References

Esch, G.W. 1999. Musings of a mentor: Acceptance of the Clark P. Read Award. *Journal of Parasitology* **85**: 1008–1010.

Esch, G.W., W.E. Johnson, and J.R. Coggins. 1975. Population biology of *Proteocephalus ambloplitis* in the smallmouth bass. A commemorative volume of the Oklahoma Academy of Science published in honor of Professor J. Teague Self. *Proceedings of the Oklahoma Academy of Science* **55**: 122–127.

Eure, H. 1974. Studies on the effects of thermal effluent on the population dynamics of helminth parasites of the largemouth bass, Micropterus salmoides. Ph.D. dissertation, Wake Forest University, Winston-Salem, North Carolina, 95 p.

Goater, T.M., C. Goater, and G.W. Esch. 2014. Parasitism: The diversity and ecology of animal parasites, 2nd Edition. Cambridge University Press, Cambridge, 497 p.

1 The Beginning

The beginning is the most important part of the work.
The Republic, Book I, Plato (427–347 BC)

I admit to being a Great Depression child, born in the mid-1930s, but I do not remember very much about those years. Based on what I was to learn later, however, I am glad I did not know about what was going on at the time. It was incredibly difficult for most folks, a real understatement when it comes to my own family. I then grew up during WWII and recall a great many things about those grim years. They were also very tough times but in very different ways from the Great Depression.

Baseball, school, and jobs occupied most days during my teen years. I frequently think about those times. However, my thoughts were always mixed in with what I really desired, but could not acquire, and that was a career in my beloved sport. I do not know how far I missed my dream of a lifetime in baseball, probably a lot more than I sometimes think. As a pitcher, I never did throw hard and was horribly wild on occasion, but I did have a pretty good "hook" (curve ball to the nonbaseball reader). With baseball realistically gone, I finished high school and headed for Colorado Springs and Colorado College (CC), where I was to begin my real lifetime. I did play 1 year of baseball at CC, but physically, I could not throw a ball after my freshman year because of arm trouble.

Several very important things were to happen during my 4 years there. The most important, by far, was meeting my beautiful wife, Ann, on a blind date. Even though we both grew up in Kansas about 20 miles apart (she in Newton and I in Wichita), we did not know each other until the middle of my third year, and her second, at CC. With her, I began my adult life and eventually my career as a parasitologist.

CC was, and still is, a liberal arts institution of about 1800–1900 students. When I began there, I was given a half-tuition scholarship. Tuition at that time was something

Ecological Parasitology: Reflections on 50 Years of Research in Aquatic Ecosystems,
First Edition. Gerald W. Esch.

like $350 per year—now, it is in the neighborhood of $40,000! As a freshman, I was enrolled in a general zoology course taught by Dr. Robert M. Stabler, or "Doc," as he was known by all, students and faculty alike. I was a real rookie that first semester. In fact, at first, I could not figure out why there were so many doctors on the faculty. I honestly did not know there was such a thing as a PhD. No one in my family had gone to college, so there was no reason that I should have known there was more than one kind of doctor. In fact, neither my mother nor father had finished high school—as I alluded to earlier, the Great Depression days were tough for a lot of people!

I can still picture Doc Stabler really well. He was an East Coast native and had graduated from Swarthmore College and then studied at the University of Pennsylvania where Dr. David Wenrich, a renowned protozoologist of that era, first mentored him. After finishing at Penn, he worked with Dr. Robert Hegner at Johns Hopkins University where he received his ScD. He then taught at the University of Pennsylvania until 1947 when he moved to CC as chair of the zoology department. He changed a lot after coming West, especially his attire, which always included a good looking Western hat, cowboy boots, blue jeans with a Western belt and a large silver buckle, and a Western shirt with pearl buttons.

While lecturing that first semester, he would frequently jump up on a long dais in the large classroom on the top floor of a rather old building that served as an academic home for several departments. On the dais, Doc would "strut" back and forth as he harangued his classroom full of naive freshmen. He was quite a sight and an absolutely fantastic lecturer, which always reflected his dominating persona.

Over the front door of the old building, "Seek ye the truth, and the truth will set you free" was carved in red sandstone. The structure was named Palmer Hall, for General William Palmer, a Civil War veteran and Medal of Honor winner, who also laid out the street plans for Colorado Springs. I recall that Palmer had also started the Denver and Rio Grande Railroad and, to my knowledge, one of the first to have a north–south route instead of east and west. It ran from Denver all the way down to the Gulf coast of Texas.

When Doc arrived in Colorado Springs, he purchased about 40 acres of land on the northern side of the city. He dubbed his property the "Venom Valley Ranch," and, yes, he kept his home filled with rattlesnakes that he frequently brought to class so he could show off his prowess as a snake handler. It was rumored he had been bitten frequently enough that the next time would be his last, because of an alleged allergic response to the snake antivenom.

My goal when I first enrolled at CC was to become a physician, but after a year, I changed all that and switched to physical education thinking I would like to teach history and coach baseball in high school (my hopes regarding a baseball career had realistically dropped several notches by that time). One of the requirements by the physical education department was a course in mammalian anatomy. So, I registered for it and then fell in love with zoology again when I took the course from Dr. Mary Alice Hamilton, Doc Stabler's sister-in-law. At this point, I had my first thoughts about graduate school, so I switched my major back to zoology.

Along about that time, I also recall asking Doc Stabler why he had chosen to become an academic and not pursued something else. His answer was succinct. He replied, "Three reasons," and after a pause, came, "June, July, and August!" When I first thought about it, I believed he was being facetious, but he was not. It took me a while, but I finally figured out what he had said. There is not another profession (except the US Supreme Court and the US Congress during an election year) that allows one to have almost 3 months every year (not counting Thanksgiving, Christmas, and New Year vacations, plus a full week each spring) to do whatever you like, for example, read, do research, travel, write, or all the above or do nothing. Moreover, throughout each academic year, you are in contact with young inquisitive minds with new ideas and ways of thinking. And, you actually get paid for doing it!

When my senior year came around, I decided to take advantage of the liberal arts curriculum offered at CC. I had a very strong attraction to history and had identified a couple of courses that sounded fascinating. Stabler, as chair of the zoology department, had to approve registration for all majors. When I showed him the card with the courses I had chosen, he looked at it for the longest time and then lifted his head and asked, "Are you going to be a zoology major or a history major?" I immediately responded, "A zoology major, sir!" He stared back and responded, "OK, you are going to take my parasitology course." I answered, timidly, "Yes sir," and that was my introduction to what was to eventually become my profession. However, I should note that we compromised, because I did manage to take one of the courses, "A History of the Trans-Mississippi West." It was all about cowboys and Indians, and I loved it!

At the time, I thought my future was in anatomy. As a senior at CC, I had even applied for, and received, a teaching assistantship in their Department of Anatomy at the University of Kansas, School of Medicine. Later that academic year, Ann and I drove to Lawrence, Kansas where I had an appointment with the chair of the anatomy department. He informed me that I had to come up to Lawrence over the next summer and take a course in gross anatomy so that I could be an assistant in the medical school's gross anatomy course the next fall—I should have figured this would happen. After a long walk back to our parked car and a brief conversation, Ann and I decided that I should resign my anatomy assistantship. Gross anatomy during the hot summer in Kansas was not for me, especially at a time when air conditioning was not yet very common. The idea of spending my summer with a cadaver just was not very special.

In a quandary, the first person I went to on our return to CC was Doc Stabler. I explained to him that I wanted to switch areas and become a parasitologist. During his parasitology course, Doc was a "name dropper," so he handed me a membership roster for the American Society of Parasitologists, at which time he was the treasurer. He instructed me to give him the names of five people by the next day and that he would have letters in the mail immediately. I can recall three of them, that is, one was Ray Cable at Purdue, another was a malariologist at UCLA, and the third was J. Teague Self at the University of Oklahoma. Doc told me later that he had written good supporting letters and ended them by saying, "First come, first served." Later that week, he received a phone call from Dr. Self who offered me a teaching assistantship beginning the next fall in their Department of Zoology at the University of Oklahoma. I took it! It was the second correct step in my career—the first was obviously in agreeing to take Doc's parasitology course.

By that time at CC, I had moved off campus and was living at home. My father had died a week shy of my 16th birthday. My younger brother, Gary, had graduated from high school and decided he wanted to come to CC too. So, our mother sold our house in Wichita and our hardware business she had been managing since our father passed and moved to Colorado Springs. She then decided, at the age of 50 and without a high school degree, she would go to nursing school and become a registered practical nurse—which she did, successfully. I even ended up tutoring her in anatomy and physiology. I also remember telling her that I was not going to medical school or to the University of Kansas in anatomy, but to the University of Oklahoma to get a PhD degree in parasitology. The hurt look on her face was an awful experience, but I immediately told her that one of the leading killers in the world was malaria and that it was caused by a parasite. Her look changed immediately and she was satisfied.

There are a couple of things that I still recall about Doc's parasitology course. I remember the day he came "swaggering" into the lab carrying several long pieces of paper in his hand and smugly tossed them down on the table where several of us were working. He announced, "This is number 100." Of course, none of us knew to what he was referring, but he then explained that on the table were the page proofs of his 100th publication. I have since learned the significance of his achievement and why he was so proud of it. I go to a baseball analogy by way of an explanation. It is like getting 3000 base hits during an entire career in the major leagues or winning 300 games as a pitcher!

Another time, he began mumbling about having to say "zooooo-ology" in a play that was being presented in a nearby community theater. As it turned out, the play was "Inherit the Wind," dealing with the 1925 Scopes monkey trial in Tennessee. Doc had the part of William Jennings Bryan, one of the prosecutors of John Scopes who had been accused of teaching evolution, which was against the law in Tennessee at the time. Clarence Darrow was his legal opponent, and Doc, playing the part of Bryan, hated every bit of it. I did not see the play, but I was told (not by Doc) that he was a pretty fair thespian.

I also recall the time he lectured to us about taeniid cestodes and, in particular, a large bladder worm, which, at that time, was known as *Multiceps serialis*. (The bladder refers to the larval stage, which is in the form of a sac that contains a transudate, or fluid, and up to several hundred scolices, all attached to the inside of the bladder wall.) Doc said the larval stage, also called a coenurus, was common subcutaneously and intramuscularly in jackrabbits out on the Great Plains east of

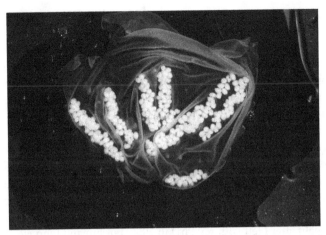

Figure 1.1   A coenurus of *Taenia multiceps*. The bladder was about twice the size of an ordinary golf ball. Note that the scolices are in several rows (hence the old name *Multiceps* [many heads] *serialis* [in a series of rows]). The coenurus was removed from a jackrabbit shot on the shortgrass prairie of eastern Colorado during my senior year at Colorado College.

Colorado Springs, with coyotes as the definitive hosts. I was very curious about such an interesting creature, and in very short order, I got out my 16-gauge, bolt-action, Mossberg shotgun and headed east, out on to the shortgrass prairie. The first jackrabbit I shot was necropsied immediately, and lo, there were three bladder worms (coenuri) ranging in size from a large walnut to a small grapefruit (Figure 1.1). All were filled with the transudate fluid, and all had many scolices attached to the inside of their walls. Little did I know at the time, but this parasite was to become a focal point for my research over the next 7 years.

I was the senior author (Esch et al., 1958) on my first paper, which was coauthored with Bob Catlett, a graduate student in the zoology department, and Dick Beidleman, a new faculty member in the Department. The "paper" was actually an abstract written for a meeting of the Colorado–Wyoming Academy of Science and dealt with an analysis of long-eared owl pellets from a ranch in eastern Colorado. I know it is a "stretch" and that most PhDs do not include abstracts as publications in their CVs. However, I have always felt that this was my first taste of being a professional biologist, and, so, I have kept it at the top of my CV throughout my career. My second publication though was legitimate (Esch et al., 1959). It was two pages long and appeared in the *Journal of Mammalogy*. However, like the first one, it had nothing to do with parasites. An old hunting buddy, Larry Long, and I, along with Dick Beidleman, had observed that breeding in the jackrabbits was early that year. The mammal people decided our paper describing the phenomenon was important enough to publish in their journal in 1959. Based on these experiences at CC, I have always felt that it was important to involve undergraduates in research, and I have tried to do so throughout my years of academic teaching. In fact, of my first 30 papers, seven were coauthored by undergraduates.

Part of the first year in graduate school at the University of Oklahoma was spent getting myself oriented. Ann and I were married on December 22, 1958, at her parents' home in Newton, Kansas. We borrowed her father's car and drove to Kansas City for our honeymoon, and then it was back to Newton for Christmas. We took a train down to Norman where we were to spend the rest of our first year in an upstairs apartment of an old house about a block from the OU campus. I recall that the train trip was very cold, and we had to sit on our suitcases outside, in between the very crowded passenger cars.

That first year, I was to also meet the rest of the students who belonged to Dr. Self and who were to become my lab mates for the next 5 years, that is, Horace Bailey, Fred Hopper, Henry Buscher, Jim McDaniel, and John Janovy Jr. I did not see the first three again after I left with my PhD in 1963, but Jim and John were to become lifetime friends, although my friendship with Jim was relatively brief because he died way too young. John married Dr. Self's secretary, Karen, and they, like Ann and me, have stayed together ever since. One of the first courses I took in graduate school was with John in ornithology, taught by George Miksch Sutton, one of the foremost ornithologists of his time. Except for having to rise every Saturday morning at 4:00 for the inexorable weekly field trip, the course was actually great fun!

It was during my first semester at Oklahoma that I encountered *M. serialis* again. In the fall of that year, Dr. Self and I sat down to discuss the research I would do for my master's thesis. He informed me that he had a large collection of taeniid cestodes, which had been assembled by a wildlife biologist (Frank B. McMurray). The hosts included gray foxes, feral house cats, and coyotes from the Wichita Mountain Wildlife Refuge (southwestern Oklahoma) collected by McMurray between 1936 and 1941. Would I be interested in undertaking a study of their taxonomy? The collection was large, on the order of nearly 2000 worms, with a 60/40 split between immature and mature taeniids, respectively. The systematics of the group was in some flux, and since I had done some work with larval *M. serialis* as an undergraduate, I jumped at the opportunity.

At the time of the master's study, *Taenia pisiformis*, as a genus and species, was safe as a taxon. However, I discovered two other taxa in the collection, namely, *Hydatigera taeniaeformis* and *M. serialis*, and the taxonomic validity of each was questionable, that is, *Hydatigera* at the generic level and *M. serialis* at both the genus and species levels. After staining the specimens and measuring a variety of scolex and proglottid features, I concluded that *Hydatigera* was inappropriate as a distinct taxon and should be synonymized with *Taenia* and that the other questionable cestode was *Taenia multiceps*. I was very flattered when I later learned that the referee for our paper (Esch and Self, 1966) was none other than the late Robert Rausch who was one of the world's experts on taeniid cestodes! I learned later that Rausch had been skeptical of the taxonomic position of *T. multiceps* but that we had convinced him regarding our conclusions.

Our assertion regarding the taxonomic status of *Taenia (M.) serialis* was to have interesting implications later. The decision was based on several morphological characteristics associated with various species of what was considered as *Multiceps*. One of the problems as I saw it then, and see it still, was that there were several species of *Multiceps* with overlapping morphometrics of structures that could not be affected by fixation techniques or the state of relaxation or contraction at the time of fixation. Of course, I am speaking about hook measurements and shapes. Moreover, there were two other features that were, and still are, evaluated when attempting to deal with the systematics of *Multiceps* spp. If one examines the morphologies of adult cestodes in species classified as *Taenia* and those of *Multiceps*, there are no real differences, except for consistent differences in their size. The second characteristic is the larval stage, which for *Multiceps* is the coenurus, as opposed to a cysticercus for most other *Taenia* species, including *T. pisiformis*. If all aspects of the biology and morphology of *Multiceps* spp. are carefully considered and compared with the same features associated with species of *Taenia*, species in the two genera are not that different, except for the coenurus versus the cysticercus, and in my opinion, that was insufficient to keep them separated at the generic level.

As stated previously, the question regarding the separation of *T. multiceps* and *T. serialis* was resolved by synonymizing *serialis* with *multiceps*, since the latter had priority. Another question persisted, however, and that had to do with the location of coenuri in their intermediate hosts. In sheep, these larvae occur in the brain where they cause "gid," or "staggers." In jackrabbits, coenuri are intramuscular or subcutaneous. As I stated in my thesis, "This difference could well be a condition of host influence on a single species and is thus not a valid species character until proved or disproved experimentally." When I wrote that sentence, I had absolutely no idea about what was to be eventually discovered down the road.

One of the things I have noted in my experience as a parasitologist, or of science in general, is the speed with which some things change. Most of this is based on technological developments that occur, which have application to both ecological and biochemical/physiological parasitology. Some may wonder about my assertion regarding ecology and technological advances. In this regard, all I have to do is say, computers and Geographic Information Systems (GIS), and that should take care of the issue.

A long about that time, in the early 1960s, some of the really big names in parasitology were Clark Read, Theodor von Brand, Moises Agosin, and Don Fairbairn, to name just a few. These folks were biochemistry/physiology types, which was then the really "hot" area for parasitology. Dr. Self felt that if I was going to make a name for myself, this is where it would come, even though he knew very little about these disciplines himself. Again, I was lucky because there was a very good biochemistry PhD student in our department at Oklahoma, Cal Beames, who was willing to help me get started with my research. Another person, Dr. MacWilson Warren, was a young faculty member at the University of Oklahoma Medical Center in Oklahoma

City where Dr. Self's lab would go every other week for a bag-lunch seminar. Our trips up there allowed me to develop a very nice relationship with Mac Warren that continued over the years. I had by then decided to stay for my PhD with Dr. Self, another really good move on my part.

During that first year of my PhD effort, I was encouraged by Dr. Self and Mac Warren to apply for a National Institutes of Health (NIH) Predoctoral Fellowship. Mac was very instrumental in his help with writing the grant proposal, and I was successful! What a difference in my life and Ann's after the NIH grant came through. By that time, we had a baby son (Craig) to take care of, and my salary went from $1350 a year (yes, $1350 a year is correct!) to $3600 per year (tax-free and a lot of money in those days!); also, it provided for a modicum of supplies and equipment and travel funding to attend a scientific meeting every year.

So, what was my choice for research? My NIH proposal was to compare the intermediary carbohydrate metabolism in a larval and an adult tapeworm. The idea was that the parenteric larvae occurred in an aerobic environment, while adults were enteric, a largely anaerobic environment. Were the metabolic pathways of our parasite different in the two places? At least, this was the basic question. And, what was my choice for an experimental model? Well, why not *T. multiceps*? Obtaining a dog and adult tapeworm proglottids was easy. I shot a jackrabbit, removed a coenurus, and fed it to a young pup—6 weeks later, it was shedding gravid proglottids in the feces every morning. I had read in the literature that laboratory rabbits could be infected with *T. multiceps*, and it should have been a cinch to obtain coenuri, but it wasn't. I could not obtain an infection of rabbits in the laboratory—it would not work. I also expected to find an abundance of eggs in gravid proglottids, but every time I opened up a shed gravid proglottid, very few eggs were present. I could not figure out why? There was a very good reason that my effort was unsuccessful, but it took me three more years until I was doing a postdoc to figure out my mistake—more on this later.

Dr. Self knew about my frustration and that it was going to be necessary for me to shoot jackrabbits and obtain coenuri in the field. So, he put me in contact with a rodent and predator control officer who worked for the US Department of the Interior. His name was Lyle G. Rexroat, and I will never forget him—what a character! His job was to kill coyotes and badgers out on ranches near a small Oklahoma town named Fort Cobb, about 90 miles due west of Norman. It was the site of an old 1880s army fort used in the days of the Indian wars on the Great Plains of the Old West.

By that time, Ann and I had purchased a new Renault automobile. The car was similar in many ways to a Volkswagen. It was small, and the trunk was in front, with a large, hard rubber plug sitting in a drain hole at the bottom. On each trip I took over to Fort Cobb, I carried my 16-gauge Mossberg shotgun, plenty of shells, dissecting tools, and 10–15 coffee jars filled with physiological saline. Leaving at 4:30 in the morning, I would arrive at around 6:00. Lyle and I would transfer everything to his old Chevy pickup truck and head out of town. He knew all of the ranchers in the area, so it was no problem for us to simply drive from one pasture to another. Interestingly, the jackrabbits were never "skittish." In fact, they would sit upright and never run from the truck as we drove through the pastures. It was easy to shoot 10–12 rabbits from the passenger side and toss the carcasses into the bed of his truck. Since the prevalence of infection was around 60%, and multiple infections were common, it was easy to garner enough coenuri to keep me going in the lab for 2–3 days, and then it was back to Fort Cobb. Removing coenuri from the rabbits was also simple with a pair of sharp scissors and a scalpel, and then into the jars with physiological saline went the coenuri. They would be placed into the trunk of my car, and two or three bags of crushed ice added to keep the larvae (and me) "comfortable" as I drove back to Norman—by that time of the morning in the summertime, the temperature could easily be 90°F. The jars containing the larval cestodes were carried to my lab and the rubber plug removed from the trunk, allowing the ice to melt and the water to drain in the hot sun.

With a constant supply of gravid proglottids in the stool of the dog every morning and easy access to a supply of coenuri, a significant part of my research was uncomplicated. Enzyme assays were also easy. I would simply purchase the assay kits from Sigma in St. Louis, Missouri, and run them using a colorimeter—nothing sophisticated. The most difficult part of the work came through the necessity of comparing oxygen consumption by the larval and adult tapeworms. I had great problems with the "Warburg" instrument (or respirometer), especially in Oklahoma and even more especially in the spring and early summer. If you have never used the technique, my advice is, don't, if you can avoid it! A Warburg respirometer is what I would call contrary, even nasty. The idea behind the instrument is relatively simple. A change in the volume of gas consumed by a piece of tissue in a small flask can be measured by changes in pressure inside the manometer. The problem comes when you are working, and everything is going well and, suddenly, along comes a thunderstorm. The thunder, lightning, wind, and hail are not the problem; after all, you are inside where it is nice and cozy. The difficulty came with the huge changes in barometric pressure that usually accompany an Oklahoma thunderstorm. If I were running a dozen manometers, by the time I had recorded the change for each manometer, I would frequently need to go back and start reading them again, immediately. And naturally, following a thunderstorm, barometric pressure would also rise rapidly, frequently as fast as it had fallen. Several times, I had to repeat experiments to make sure I was not looking at spurious results. Despite the sometimes unfavorable weather conditions, I completed my dissertation research and received my PhD degree on time.

One of the most interesting experiences during my last spring in Oklahoma came after I had finished my bench work and was in the process of writing my dissertation. I had already applied for an NIH Training Grant postdoctoral position in the Department of Parasitology, School of Public Health, at the University of North Carolina, Chapel Hill (UNC–CH). One day that spring, I received a phone call from Doc Stabler (back at CC, my alma mater). It seems that one of his senior faculty members had impregnated an undergraduate coed. That was bad enough, but to make matters as bad as they could become, she had committed suicide in the stockroom of the zoology department. Doc wanted me to come back to CC immediately and take his job. I told him that I wanted to consult with Dr. Self before making a decision. I was in a huge dilemma. I had already applied for the NIH Training Program, in a prestigious department, to work with Dr. John Larsh, a very well-known immunoparasitologist of his era. However, I had not completely written my dissertation yet, and the job offer would require me to stop everything and head back to Colorado Springs, immediately. When I spoke with Dr. Self, he pointed out that going back to CC as a junior faculty member would not be the same as when I was a student there, kind of like "you can never go home again." Moreover, the teaching load would be heavy, and pressure would be great. Finally, Dr. Self asked, "How will you know about research in your career if you do not experience it first?" That settled the issue. I called Doc Stabler back and told him that I simply could not do it. He was not very understanding, commenting, "Do you want to be a baby and do a postdoc?" or words to that effect! Ironically, I recently turned down a position at CC a second time, in May 2012. They wanted me to chair their Department of Biology for 3–4 years, which meant that I would need to retire from Wake Forest in order to take the job. I agreed to take a 1-year appointment, but they rejected my counteroffer. This time, however, we parted amicably.

So, on completing my dissertation, successfully defending it, and publishing it (Esch, 1964), Ann and I headed to North Carolina, along with our two very young children (our daughter, Lisa, came in the spring of 1962). We promised ourselves that we would return to the West as soon as our "tour of duty" was over after 2 years in Chapel Hill—how little did we know!

Chapel Hill is in the north-central part of North Carolina, forming one point of a triangle with the cities of Durham (Duke University) and Raleigh (North Carolina State University) as the other two points. At that time UNC–CH was not huge, probably in the neighborhood of 20,000 students when you add the graduate school and professional schools of medicine, law, dentistry, pharmacy, and so on. The Department of Parasitology occupied a floor, or most of one, in a relatively new building that housed the School of Public Health. For the most part, it was a very good place to work. John Larsh was the head of the department. Jim Hendricks, Hilton Goulson, and Norman Weatherly comprised the faculty. I was the only postdoc, but there was also a wonderful cadre of really good graduate students that included Bruce Lang, Larry Gleason, and Darwin Murrell. When Bruce finished his degree, he was to go off to Eastern Washington University (Cheney, Washington), and I have not seen him since. Larry was to become a great friend and a faculty member at Western Kentucky University (Bowling Green, Kentucky), but like Jim

McDaniel, he died very young, also of a heart attack. I have kept in touch with Darwin over the years, and we have developed an even more solid and friendly relationship.

My research at UNC–CH started off rather poorly. The first thing I decided was that I was not an immunologist and, frankly, did not want to become one either! Their research model was *Trichinella spiralis*, although Bruce managed to do some interesting immunology work with *Fasciola hepatica* in mice, a "neat" trick considering the size of the parasite and the mouse host. Since I was definitely not an immunologist, I made a pitiful effort to do some biochemistry with *T. spiralis*, but was not very successful (it was my first step in learning that I would never be successful as a biochemist). However, I really wanted to do something more with *T. multiceps*, so I convinced Larsh that he should let me ship my infected dog back east so that I could focus attention on "my" cestode. He consented, although I think reluctantly, until I later hit the research jackpot with the parasite. I was not quite sure what kind of research I wanted to do, but the dog came east, and real success was to soon follow.

Ever since I examined the first gravid proglottid in Oklahoma 3 years before, I had noticed that a milky secretion came out of one end of the proglottid, not the gonopore, but one of the attachment points of the proglottid in the strobila. At the time, I thought it was fluid from the osmoregulatory canals that was coagulating in the physiological saline in which I had placed the proglottids. As I noted earlier, one of the problems I had with infecting rabbits in the lab was the scarcity of eggs that should have been present inside each proglottid. A few were always present, but not many. Not long after arriving in Chapel Hill, I was sitting in my office one day, examining proglottids I had isolated from a dog's stool. One of the graduate students (I cannot recall which one) was watching and inquired about the milky secretion. He asked if I had ever checked the supposed "coagulant" with a microscope and I replied that I had not. He gave me a quizzical look, like why not? So, I picked up a Pasteur pipette, pulled up some of the fluid, placed it on a slide, and added a coverslip. When I looked at it using a compound microscope, I almost fell out of my chair—there were hundreds, if not several thousands of my scarce eggs. Inside the striated shell of the eggs, oncospheres were vigorously stretching and relaxing! At last (thanks to a little help from a friend), I had found the eggs for which I had been searching the previous 3 years. Before then, and up to the present time, I have never seen any reference in the literature to this kind of egg release. I had always assumed that when most cyclophyllidean proglottids were shed, they simply dried up and disintegrated, releasing their eggs in the process. I still do not know if the egg-release mechanism is unique to *T. multiceps* or if it is a characteristic of taeniid cestodes in general.

At least now I had an egg source, but what to do with them was the next question? I did not want try to infect rabbits because of my failure to get an infection back in Oklahoma. I was lamenting about this to Bruce Lang one day, and he suggested that I try a lab mouse. I immediately responded by saying that this would not work because I had never seen a reference regarding coenuri in a rodent, of any kind. He then suggested that I try shooting my mice with copious quantities of cortisone before exposing them to eggs. The idea was to reduce the inflammatory response to enhance the chance for the parasite to become established and develop. I had nothing to lose, so, that is what I did—I inoculated each mouse with a load of cortisone and intubated it with 1500–2000 eggs. Several weeks later, I noted a large subcutaneous "knot" on the neck of one of the mice. I was heading off to a parasitology meeting in Chicago and figured I would kill and necropsy the mouse when I returned. I honestly was not thinking about a coenurus. The colony of mice that was being used in the department was prone to acquire subcutaneous tumors, and I had seen them before. On returning from my trip, I isolated the mouse and killed it by cervical dislocation. When I opened the skin over the bulging "knot," I was flabbergasted to see a coenurus, just like those I had observed so many times in Colorado and Oklahoma jackrabbits. Scolices attached to coenurus wall numbered about 150. So, I approached Larsh again and asked his permission to acquire another dog, and he agreed. I fed the coenurus to the animal, and about 6 weeks later, it was shedding proglottids. Eggs were isolated from these proglottids and fed to mice using the same protocol I had used the first time, and it worked again. In following this protocol, I had also satisfied one of the so-called Koch's postulates, that is, "The microorganism must be re-isolated from the inoculated, diseased experimental host and identified as being identical to the original specific causative agent."

Something else very peculiar also happened, something that I certainly had not anticipated. Following egg intubations, I had been tracking mouse body weights. I noted that several had begun to lose weight rather rapidly. Moreover, these same mice had begun to act rather strangely. If the lid on a cage were tapped gently, they would respond by jumping, high, even to the point of banging their heads on the lid of the cage. The same mice began running in circles, like they had gid. One of these mice had died in the cage and was partially consumed by its cage mates. I removed the dead mouse and cut open the skull. Buried in the cerebrum was a developing coenurus, about the size of a pea!

This observation sort of topped things out with *T. multiceps*, for a couple of reasons. First, I knew that the parasite exhibited a tissue tropism of one kind in sheep (the brain) and another kind in jackrabbits (subcutaneous and intramuscular tissues). But now, there was a combination in the brains and subcutaneous tissues in the laboratory mice. Second, I learned something else about handling this particular parasite. I had seen in the literature several papers reporting the parasite in the brains of humans, primarily in younger children but still in the brain. In fact, the Centers for Disease Control (CDC) in Atlanta actually called me about the case of a coenurus occurring in a young lad from Wyoming and how it should be treated.

What these findings meant was that if humans ingested eggs of the parasite accidentally, they could then develop cerebral coenuri. It also meant that for a little more than 4 years, I had irresponsibly identified the nature of the secretion from gravid proglottids. During this period, I never wore plastic or rubber gloves. This, of course, begged the question, "How many (if any) eggs had I accidentally ingested because of my disdain for caution?" After the experience, I developed a very serious case of "intern's syndrome." Every time I developed a headache, I wondered about the cause. Many times in my last year at UNC–CH, I can vividly recall coming home after work and trying to see if I could stand on one leg or touch my nose with my index finger, both tests being performed with my eyes closed. However, I thankfully never did exhibit signs or symptoms of cerebral coenuri.

The success with research in Chapel Hill yielded four publications (Esch, 1964; Larsh et al., 1964; Race et al., 1965; Esch et al., 1966). In two of the four papers, I was relegated to the third spot among the three authors. I remember the circumstances regarding authorship "assignments" quite well. I was called in to visit with Larsh. During our conversation, he indicated that, "It wouldn't look too good if he was in the third spot on two of the papers," both of which were published in the *Journal of Parasitology*. Larsh was to be the senior author on the one dealing with pathology induced by *T. multiceps*, with Race as the senior author of the one dealing with electron microscopy of the coenurus. He approved my being senior author on a paper that was focused on electron microscopy of the adult stage of *T. multiceps* and another dealing with the research protocols involved in infecting the mice with eggs of *T. multiceps*. Both of these papers were to be published in the *Journal of the Elisha Mitchell Scientific Society*.

I tell this latter story for a reason. Certainly, I was greatly angered with Larsh because I felt that I owned the research, which I did. What bothered me most was that neither Larsh nor Race "needed" to be a senior author. They were both already well established, and a couple of years later Larsh would actually be elected president of the American Society of Tropical Medicine. It should be added here that when Dr. Self got wind of the situation, he told me not to be concerned about it and that he would make sure that the "powers that were" in the American Society of Parasitologists would know about it—and he did! Because of this rather painful episode, I decided that if I was ever in a similar position with a graduate student or a postdoc and it was clearly their research to be published, I would make no demands regarding seniority. I have honored this self-pledge ever since, and I am proud in having done so. It is an absolute that graduate students must receive full credit for the research they accomplish no matter the intellectual contribution by a mentor.

There is one more important thing that I recall about my experience at UNC–CH. During my 2 years there, William Walter Cort, who had retired to Chapel Hill, was a frequent visitor. He would come to my office and spend an hour or two where he would regale me with stories about his career. One of the things I really liked about this man was that he had graduated from CC in 1910 and had played varsity football on a team that had beaten the University of Colorado twice during his collegiate career. Following CC, he attended the University of Illinois where he worked under Henry Baldwin Ward for his PhD. Ward was not only the founder of the *Journal of Parasitology*; he is considered by many to be the father of parasitology in North America.

Cort was but one of many, for example, Standard, LaRue, Thomas, and so on, of the early and great parasitologists who had passed through Ward's lab. After teaching for a few years in the Midwest and on the West Coast, Cort settled in at Johns Hopkins University in Baltimore, Maryland, where he began a long-lasting reign, not unlike that of Ward. His research had considerable breadth. He was well known for his work on hookworms, and anyone who has done any aquatic research in the upper Midwest of the United States will certainly know about swimmer's itch, the cause of which was discovered by Cort while working at the University of Michigan's biological station in the 1920s.

When I came to Wake Forest in 1965, I played around with *T. multiceps* a little while longer. However, I soon switched parasites along about the time Ray Kuhn joined our faculty from the University of Tennessee (Knoxville, Tennessee). Ray and I coupled with our research in 1967, and I was very fortunate to collaborate with him over the next 3–4 years. He was trained in developmental biology, and I had an interest in the developmental aspects of larval cestodes, so it became a natural partnership.

Reino (Ray) Freeman of the University of Toronto had isolated what he called the Ontario Research Foundation (ORF) strain of *Taenia crassiceps* (Figure 1.2) in 1952 (Freeman, 1962). The definitive hosts for the parasite are foxes and the intermediate hosts are voles, *Microtus pennsylvanicus*. In the normal intermediate host, larval cysticerci of *T. crassiceps* occur in subcutaneous tissues where they bud exogenously, producing more cysticerci. When Freeman began his isolation work on *T. crassiceps*, he at first maintained the parasite by passage through ordinary lab mice and domesticated dogs, and then he used eggs from gravid proglottids to infect new mice. Subsequently though, he bypassed the canine host by inoculating new hosts intraperitoneally with small larval buds that had just been disengaged from the parent. Later, however, when he attempted to infect dogs with cysticerci from mice, the infection would not take. Moreover, by 1965, he observed that the scolices on the ORF strain possessed irregularly shaped rostellar hooks. We obtained the ORF strain of *T. crassiceps* sometime between 1965 and 1969. By 1969, hooks were no longer associated with *T. crassiceps* cysticerci, and scolices, even partially developed ones, were rare. In other words, larval ORF cysticerci had become anomalies, which, in the wild, would have disappeared in a single life cycle generation.

In 1965, Fran Dorais, a high school teacher in Lincoln, Nebraska, came to work in my lab in pursuit of his master's degree at Michigan State University when I was teaching in the summers at the Kellogg Biological Station (KBS). That summer, he isolated cysticerci of *T. crassiceps* from a vole (*Microtus* sp.) he had trapped near the station. Passage from mouse to mouse of what we called the KBS strain was initiated using Freeman's intraperitoneal inoculation technique. That first summer, Fran and

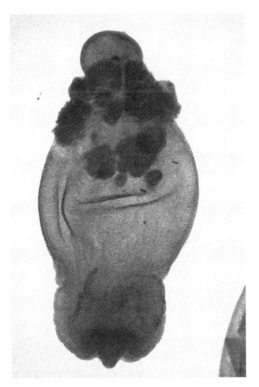

Figure 1.2   A cysticercus of *Taenia crassiceps*, about 5 mm in size, from top to bottom. This cysticercus belongs to the anomalous Ontario Research Foundation (ORF) strain isolated by Reino (Ray) Freeman at the University of Toronto. The scolex development has begun at the top, but it would not have proceeded any further than what can be seen here. Exogenous budding is occurring at the other end.

I designed an experiment to compare growth rates of the ORF and KBS strains of larval *T. crassiceps* (Dorais and Esch, 1969). We found that not only did the ORF strain reproduce more rapidly, but they also possess more buds per larva; in other words, they had become reproductive machines, with a syringe in place of a canine host. Later, in another study, Ken Smith (an undergraduate student), Ray Kuhn, and I (Smith et al., 1972) reported that the anomalous reproductive biology and morphology in the ORF strain could be accounted for by aneuploidy. The diploid number was 16 for the KBS strain and 14 for the ORF larvae. On examining the karyotypes of the two strains, we found that a single pair of chromosomes were missing from the ORF larvae and, accordingly, reasoned that the absence of these chromosomes was responsible for the unusual growth rate and scolex malformation.

It has long been known that proteins of considerable molecular weight can accumulate in the bladder fluid of various taeniid cestodes, including *T. multiceps*.

In fact, the fluid is referred to as a transudate of host serum and carries common plasma enzymes, for example, phosphohexose isomerase, aldolase, and lactic dehydrogenase, as well as other proteins (Esch, 1964; Chordi and Kagan, 1965). However, the exact source of these proteins in bladder fluid remained unknown for a period of time before Ray Kuhn and I decided to determine if, in fact, large, what I would call "foreign," proteins could enter the bladder transudate from outside. One of the interesting aspects of this particular research project, but not really germane to the science we performed, was that we actually did all of the experimental work during a single weekend!

Our experimental design was simple. We used ^{14}C-*Chlorella* protein as the foreign protein source. Our results clearly suggested that the protein was taken up whole and that it was not metabolized in any way by the parasite while it crossed the tegument and the underlying tissues of the bladder wall (Esch and Kuhn, 1971). The question then became, how does it get across? We suggested an idea based on some observations of Bob Morcock and Phil Mount, a couple of graduate students working in my lab at the time. Each of these guys had seen what they described as 3–5 μm diameter canals in the walls of the *T. crassiceps* bladders, which would provide easy access for the large proteins into the bladder fluid. However, we never followed up on this idea.

Examination of the tegument of cestodes reveals a highly complex structure, with a microtrich (microvilli) border on the outside, small vesicles just inside, cristate mitochondria along a basal membrane, and periodic cytons (cell bodies containing typical organelles) extending below the basal membrane. The cytoplasm of the cytons (and their nuclei) is continuous with the tegumental cytoplasm, resulting in what is referred to as a syncitial organization of the cestode surface. The vesicles have long been thought to be the result of pinocytosis (phagocytosis on a small scale). Ambrosio et al. (1994) reported evidence for what they termed "adsorptive endocytosis" by larval *T. crassiceps*. They even generated data supporting the notion of "degradation of internalized proteins" and suggested the possible use of these macromolecules as nutrients for the parasite.

The *T. crassiceps* work that Ray, our students, and I completed represents a strong collaborative interaction. Despite our success, it was at this point that Ray and I separated ways with respect to our research. We did so for a couple of reasons. While a graduate student at the University of Tennessee, Ray had taken a techniques course dealing with developmental biology in chick embryos and had developed a strong interest in understanding the immunological aspects of this process. In the summer of 1972, he, along with a hundred or so other biologists, attended a 2-week course in immunology, at the Scripps Institute in La Jolla, California, sponsored by the American Association of Immunology. When he returned to Wake Forest, he thought initially about using *T. crassiceps* as a model for his research but then realized a protozoan parasite would serve his interests better. In addition to being a protozoan, the parasite had to be one that he could easily count and capable of using a mouse as host. His choice was *Trypanosoma cruzi*. He was off and running as an immunoparasitologist, a very good one at that!

I was also about to make a switch in research emphasis, but it took me a while longer because I had an excursion that I wanted to pursue first.

References

Ambrosio, J., A. Landa, M.T. Merchant, and J.P. Laclette. 1994. Protein uptake by cysticerci of *Taenia crassiceps*. *Archives of Medical Research* **25**: 325–330.

Chordi, A., and I.G. Kagan. 1965. Identification and characterization of antigenic components of sheep hydatid fluid by immunoelectrophoresis. *Journal of Parasitology* **51**: 63–71.

Dorais, F.J., and G.W. Esch. 1969. Growth rate of two *Taenia crassiceps* strains. *Experimental Parasitology* **25**: 395–398.

Esch, G.W. 1964. The effects of cortisone and pre-infection starvation on the establishment of larval *Multiceps serialis* in mice. *Journal of the Elisha Mitchell Scientific Society* **80**: 114–120.

Esch, G.W., and R.E. Kuhn. 1971. Uptake of ^{14}C-*Chlorella* protein by larval *Taenia crassiceps*. *Parasitology* **62**: 27–29.

Esch, G.W., and J.T. Self. 1966. A critical study of the taxonomy of *Taenia pisiformis* Block, 1780; *Multiceps serialis* Leske, 1780; and *Hydatigera taeniaeformis* Batsch, 1786. *Journal of Parasitology* **52**: 932–937.

Esch, G.W., R.H. Catlett, and R.G. Beidleman. 1958. An analysis of long-eared owl pellets from northern Colorado. *Journal of the Colorado-Wyoming Academy of Science* **4**: 49.

Esch, G.W., R.G. Beidleman, and L.E. Long. 1959. Early breeding of the black-tail jackrabbit in southeastern Colorado. *Journal of Mammalogy* **40**: 442–443.

Esch, G.W., G.J. Race, J.E. Larsh, and J.H. Martin. 1966. A study of the adult stage of *Taenia multiceps* (*Multiceps serialis*) by electron microscopy. *Journal of the Elisha Mitchell Scientific Society* **82**: 44–56.

Freeman, R.S. 1962. Studies on the biology of *Taenia crassiceps* (Zeder, 1800) Rudolphi (1810) Cestoda. *Canadian Journal of Zoology* **40**: 969–990.

Larsh, J.E., G.W. Race, and G.W. Esch. 1964. A histopathology study of mice infected with the larval stage of *Multiceps serialis*. *Journal of Parasitology* **51**: 364–369.

Race, G.J., J.E. Larsh, and G.W. Esch. 1965. A study of the larval stage of *Multiceps serialis* by electron microscopy. *Journal of Parasitology* **51**: 45–52.

Smith, J.K., G.W. Esch, and R.E. Kuhn. 1972. Growth and development of larval *Taenia crassiceps* I. Aneuplody in the anomalous ORF strain. *International Journal for Parasitology* **2**: 261–263.

2 The End of the Beginning

The great law of culture is: Let each become all that he was created capable of being.
Richter, Thomas Carlyle (1795–1881)

In the spring of 1965 and my last semester as a postdoc at the University of North Carolina–Chapel Hill, I began job hunting. I found the description in *Science* magazine of a position open at what was then known as Wake Forest College in Winston–Salem, North Carolina. Having spent a year and half down the road from Winston–Salem, I knew something about the school, but not much. I applied anyway and was invited for an interview. On arrival after the short drive over from Chapel Hill, I discovered an academic community that included about 3800 undergraduates and a small graduate program. The biology department included 12 faculty members, 1 secretary, and approximately 20 graduate students. My interview took roughly half a day, and a seminar was not required.

When they offered the job, I took it. However, during my first two years at Wake Forest, I became afraid that I would die professionally there. At the time, leadership at the very top was lacking. Harold Tribble was president; he had just moved the college in 1956 from old Wake Forest (a small town about 20 miles east of Raleigh, North Carolina), and I honestly believe that he was just plain worn out. After all, moving an entire campus a hundred miles and overseeing construction of a brand new physical plant would "take the starch out" of anyone.

So, I began searching again and interviewed at the University of Pittsburg and University of Tennessee, both of which offered jobs. To this day, I am not sure why I turned them down, but I did. I also was subsequently offered a position at the University of Georgia, but turned that one down as well.

Then, in 1967, Dr. James Ralph Scales became the new president at Wake Forest and everything became very different quickly. His first action was to change the

Ecological Parasitology: Reflections on 50 Years of Research in Aquatic Ecosystems,
First Edition. Gerald W. Esch.

name of the college to Wake Forest University, with the undergraduate component designated as the college. The university name was an umbrella that included the college, the schools of business, law, divinity, medicine, and a graduate school.

Over the next 45 years or so, almost everything I now remember about the college at the beginning of my career here was either eliminated or radically transformed. Not long after President Scales realigned the college with its other academic counterparts in the university, Dr. Elton Cocke stepped down as chairman of our department. Dr. Cocke's title was chairman, but he was really the head when I arrived. What is the difference between a head and a chairman? The head is an autocrat who runs a department. A chairman has a great many of the same administrative responsibilities as a head, but the department operates democratically on most issues. Dr. Cocke represented the "old school" in terms of running a department, for example, no committees, no voting and so on. By the way, I have absolutely no contempt for the man; just the opposite, he was a truly wonderful person. In fact, I have the utmost respect for him, especially because he knew it was time for him to step aside!

Following Dr. Cocke's resignation, two important events were to then take place in our department. First, Ralph Amen was asked to become the new chairman and he really turned everything around. In fact, many of the things we do today in our Department of Biology began with Ralph. I fully credit him with making things move in the right direction. Second, with Ralph's leadership, the biology department acquired a PhD program in 1969. We were the first of the three departments in the college to achieve this goal. Between then and now, our department has grown to include 21 faculty members, 7 staff, 2 secretaries, and about 35 graduate students and postdocs. The space we occupy as a department doubled. Seven new classroom, research, and administration buildings have appeared across the campus since 1965, and six new dorms to house about a thousand new students have been constructed. Over the past several years, *U.S. News & World Report* (a national news magazine) has consistently ranked Wake Forest among the top 30 universities in the United States. In my opinion, there are three things that are required to achieve this kind of success. The first is a "committed-to-excellence" administration (from the trustees down to the deans). The second is a quality faculty (with full encouragement, freedom, and support from the administration to teach effectively and do research). Finally, we must be able to recruit an intelligent and enthusiastic group of graduate and undergraduate students, which we have done.

I now know why I did not leave—it took 48 years to get where we are, but it has been worth seeing it happen and to have been part of the transition from that small, regional liberal arts college I first knew to a major national university! In retrospect, I made the right decisions, first to come here and then to stay.

Unfortunately, while writing this book, my good friend Ralph Amen died after suffering through a long illness. He was 86 years old. Ralph was not only a good biologist, but he was a very insightful man. Though not a parasitologist, I always took whatever I wrote to him and asked for his opinion, for example, was the English clear, were the ideas any good, did I "overwrite" a sentence, and so on. Even though I began writing the book more than 2 years ago, unfortunately, he was not able to give me any input this time. His illness had taken a lot of "sap" out of him, and he will be sorely missed. His passing is also why I dedicated this book to him.

In 1970, I decided that if I was to continue my biochemical and physiological excursion as a parasitologist, I must learn how to do in vitro culture. The best person in this field was J. Desmond Smyth at the National University in Canberra, Australia. At an international parasitology meeting in Washington, DC, in 1970, I cornered Dr. Smyth and inquired about the possibility of working with him during a sabbatical I was planning to take from Wake Forest. He agreed to have me but suggested that I postpone it a year. He was moving into a chair at the Imperial College of Science and Technology in London the next fall, and he felt that my family and I would be better off in the United Kingdom than in "Oz," down under. So, we waited a year.

Desmond had learned his research craft at Trinity College in Dublin, where he had begun his culture studies using *Ligula intestinalis* as his first model. While he was in Canberra, however, he had developed a highly successful in vitro technique for *Echinococcus granulosus*, which I felt was perfect for me since I wanted to culture *Taenia crassiceps*. Because both were taeniid cestodes, I naively (and incorrectly—at least in part) reasoned that if his in vitro culture worked for one taeniid species, it ought to work for another as well. As it turned out, I was correct in selecting my new research model, but when Desmond was to return to the lab in London, he was to have some real problems with his old one.

Smyth was a gentleman, with a very slight Irish brogue, a glint in his eye, a very positive demeanor, and a great sense of humor. I was so fortunate to be able to work with him over a 9-month period. To support our trip, I managed to secure a World Health Organization Research Fellowship, along with some financial help from Wake Forest. I felt we were in a perfect fiscal position to make it through the 1971–1972 academic year.

So, in late August of 1971, Ann, our three children (in addition to Craig and Lisa, Charlie was born in 1967), and I boarded a United Airlines flight from Colorado Springs, and we were off to London via Chicago. We arrived at Heathrow and immediately made our way to the Eden Hotel in South Kensington, where we all "flopped" and tried to recover from jet lag. It took us a couple of days to make our way back into the real world. The first thing we had to do was find a place to live. We knew absolutely nothing about London, for example, how large and diverse it was, or the underground transportation system (the tube), or the funny looking taxis, or driving on the "wrong" side of the road, or the huge double-decker buses, or the fascinating pubs, or their strange accents, or that sometimes London hotels had bathrooms that had to be shared by several hotel guests, and so on, but we learned quickly!

Soon after recovering from the bad case of jet lag, Ann took the kids with her and went in one direction searching for an apartment and I went another. We were, indeed, very, very lucky. Within a couple of days, she had found a modern, three-story walk-up in a village called Ham, to the south and west of Kensington, the location of Imperial College. In the United States, our new home would be described as a "semidetached house," or a condo. It was about a 20 minute bus ride south of Richmond on Thames, where I could catch the District Line underground, pass by Kew Gardens, cross the river, and be at my tube stop, Gloucester Road, in about 30 minutes. Then, I needed to walk about five blocks, and I was in the zoology building on the campus of Imperial College.

I was able to contact the owner of the property by telephone regarding our potential temporary home, but he was reluctant to say yes when I offered to sign a lease because we wanted it for just 9 months. When he and I spoke, he told me that he had just come from playing a round of golf. Before he could say anything more, I mentioned that Arnold Palmer was a graduate of Wake Forest University (actually, Palmer had been a student at Wake, but he never graduated), where I taught in the biology department. The "edge" on his voice softened immediately, and he told us to go to Harrods Real Estate (yes, of Harrods Department Store!), where we could sign our lease, and so on.

Looking out of the upstairs back bedroom of our temporary home, the trees lining the Thames were in plain sight, and there were no buildings of any kind separating the river and us. We were also very close by the Teddington Locks, which were about a quarter mile distant. On many mornings, I can vividly recall seeing thick ground fog between the river and us, but you always knew the river's location because of the trees reaching above the fog. I frequently envisioned Henry VIII and his entourage on horseback, riding through the fog after spending the early morning hunting deer or wild boar, on their way back to Richmond Castle, which was fairly close by. Also, above the Thames, I would frequently see a fairly large bird that hovered above the trees—it did not seem to fly in one direction or another, but like a kite, it seemed to stay in one spot. I decided that it was either an osprey or a kestrel, which, as it hunts, seems to stay in one spot by flying against the wind.

About a block away from our new home and across the street from my bus stop, there was a pub, The Water Gypsies. It was to become my first stop after leaving the bus from Richmond on my way home. Eventually, it became "my" pub, where I would halt each evening and have a pint or two of bitter (later, lager) while sitting and reading the *International Herald Tribune* so I could keep track of what was going on back home. Ann would usually join me on Friday evenings. We would eat some "pub grub" and enjoy a band that was fond of playing and singing American folk music. It was great fun, but I must state that listening to the "The Wabash Cannon Ball" being sung in "real" English was a completely new experience for both Ann and me.

Once we were settled in our temporary home, things became easier. The older two children were enrolled in English schools and seemed to take to them relatively well. London was an exciting place to spend a year. It was during this time that

Parliament was preparing to vote on whether to join the European Union or not—they did, although with some reluctance. As it turns out, Ann was taking a bus tour with our two older children the evening of the vote (while I was at home baby-sitting the youngest of our three children). They were close by Westminster Abbey and the Palace of Westminster (their Parliament building) when a large crowd that had gathered to celebrate the EU membership event swallowed them up. We were also lucky because that year, the United Kingdom had changed from the old "shilling" system to decimal currency, which made it so much easier for visitors like us to understand how to pay for a "pint of bitter" at the pub.

Unfortunately, we were unable to travel outside London very often, but Ann and I have made up for that situation in later years. Since that first trip was over, we have returned at least a dozen times (with three wonderful trips during Christmas and the New Year Holiday) and expanded our travel horizons to Scotland, Wales, Ireland, plus a couple of excursions onto the continent. We even have our "own" hotel, the Clearlake, in Kensington where we stay each time we visit London. The hotel is not fancy, but it is very reasonable in cost and even comes with a kitchenette where we are able to cook our breakfast and dinner, all for £60 a night, permitting us to do, and see, a lot more than we could otherwise.

We have also come to know the couple who own the hotel quite well over the years. By birth, the husband is a Rumanian Holocaust survivor and the wife is English. I had forgotten that the Rumanians fought alongside the Germans until 1944, when Rumania signed a peace treaty with the Soviet Union and attempted to switch sides. At that point, the Germans occupied Rumania and the Nazis immediately began rounding up Jews who had been left pretty much alone during the first part of the war. At the age of 12 (apparently, he could pass for 16 quite easily), our friend was captured by the Nazis and sent to a camp in Poland where V-2 rockets were assembled. Toward the very end of the war, he escaped but was fortunately "captured" by the Americans. Eventually, he made his way to London, entered the real estate business, met his wife and married, and settled down. (Incidentally, George Soros, the liberal American entrepreneur and billionaire, was his roommate for about a year when he first arrived in London after the war.)

Now, why did I come to London and Imperial College—as I discussed earlier, it was to learn how to do in vitro culture from the "master." The work at Imperial College began slowly because I had absolutely no experience with sterile procedures and other in vitro techniques. Accordingly, I had to learn how to package and sterilize pipettes, to work with a laminar flow sterile hood, to use an inverted microscope, and so on. None of this was "rocket science," but I had to learn how to do these things before Desmond let me get started doing the culture work. As mentioned earlier, the in vitro culture system I employed was identical to the one he developed for *E. granulosus* in Australia. It was biphasic and included coagulated bovine serum as the solid phase; the liquid phase was called "Medium V." It was a mixture of Difco medium "858," heat-inactivated fetal calf serum, powdered yeast extract, glucose,

KCl, penicillin, and streptomycin. As the year progressed, we attempted several variations, but Medium V was always chosen otherwise.

After a few weeks of learning the basics, Desmond came in one morning and said, "It is time to get you started, so let's go," and we did. Larval *T. crassiceps* of the Toi (isolated from a dog by Ray Freeman), KBS, and ORF strains were employed. I was really kind of surprised when both the Toi and KBS strains worked so well. In fact, we were able to grow semiadult *T. crassiceps* the first time we tried to culture the cestode. Our worms strobilated within 96 hours, and testes began to develop by 11 days of culture. However, testes never developed further than the 64-cell stage even after 80 days in culture (Esch and Smyth, 1975). I thought it would be much more difficult and that I would probably need to make some adjustments to achieve our goal. I was, needless to say, sort of "puffed up" with our success, but I was also somewhat disappointed. I had always viewed in vitro culture technology like it was a form of alchemy. Traditional alchemy involves medieval chemistry and philosophy designed to convert base metals into gold. Obviously, this was not our objective, but the idea of in vitro culture, to me at least, should have paralleled the approach used in alchemy. The procedure turned out to be just like "cookbook" biology, with which I was already quite familiar.

Even with our success, the culturing process became tedious at times. I had to change the medium every other day in each culture flask, which meant that on either Saturday or Sunday, I was required to trek back to the lab to perform this task. Moreover, my normal weekday bus trip schedule and route were both different on the weekend, and I had to walk quite a long way to catch it on Saturdays or Sundays. To make matters even worse, the unions that included bus and underground drivers decided they would "work to rule." What that meant was, for example, if a light bulb in the vehicle had burned out, the drivers would not work until it was replaced. This caused all sorts of work stoppages, and one never knew when these delays would occur. Then, the unions that were involved in providing electricity went on strike. This meant that all electrical power would be rationed, that is, on for 3 hours and then off for 3 hours. Darkness at home in the evenings would bring out the candles. To occupy our time, I remember getting out the playing cards and teaching the older two children how to play poker. They became pretty good at it too—we had some fairly wild games because when it was their time to deal, they not only liked to make deuces wild, but they would add another card or two so that five of a kind was not only possible, it was likely. I can recall our 9-year-old daughter Lisa saying, "I don't have good cards this time, so I am flushing!" This was a very strange time in many ways, but it did draw the family closer.

When Desmond came to Imperial College in 1971, he brought with him the duties associated with editing the *International Journal for Parasitology* for which he was the founding editor. Having recently been editor of the *Journal of Parasitology* for 19 years, I have great empathy for the workload with which he was dealing at the time. In fact, except for the hours he spent teaching me the in vitro lab "tricks of the trade," I rarely saw him, at least for the first few months. Then, one day in January as I recall, he walked into the lab and announced that he was going to begin some lab work the next day. On schedule, he arrived with a liver secured at a local abattoir. In one of the lobes, still encysted in situ, I viewed my first hydatid cyst, about the size of a grapefruit. Working inside a laminar flow hood, Desmond carefully removed about half of the hydatid fluid with a needle and syringe. He then opened the bladder with a scalpel and invited me to have a look. I was astounded by the sight of so many brood capsules and hundreds of protoscolices moving with incredible vigor. Using another syringe, he began transferring small numbers of protoscolices to culture flasks and placing them into a water bath warmed to 37°C. By 6 days, the protoscolices were at what he called stage three, that is, calcareous corpuscles had almost completely disappeared, and clearly definable excretory canals were visible. However, at this stage, growth stopped. The parasites were still very active with respect to their movement, but no further development occurred in the flasks.

Obviously, Desmond was puzzled. So was I, especially since my *T. crassiceps* continued to grow successfully, just like it did the first time I tried his culture technique. After a few more attempts, he began to wonder what had gone wrong. Was it the water, the source of bovine serum, and so on? While I was in his lab until late May of 1972, I observed Desmond and Zena Davies, his excellent technician, try over and over to culture *E. granulosus* but always without success.

Two years after I completed my sabbatical, they (Smyth and Davies, 1974a) published a three-page research note in the *International Journal for Parasitology*, in which they announced a very surprising resolution for their problem. In the paper, they suggested to the parasitology community the existence of physiological "strains" of *E. granulosus*. While in Australia, Desmond had developed his in vitro procedure using hydatid cysts isolated from sheep. When he attempted to use the same technique for hydatid cyst protoscolices isolated from horses in London, it would not work, even though, ironically, it continued to work very nicely for me with *T. crassiceps*. I use the word, "ironically," because if one had to predict, success with horse protoscolices should have been achieved before the cysticerci of *T. crassiceps*.

Desmond knew that hydatid cysts occurred in UK sheep, so he obtained larvae from both sheep and horses and attempted to grow protoscolices using precisely the same in vitro conditions that he had employed in Australia. The sheep protoscolices developed just like they did in Australia, but those from horses in England continued to just "sit" there. There was consistent development to stage three by the protoscolices from horses, but as I described earlier, they stopped growing at that point. I have often wondered what my year would have been like if I had not been successful or, perhaps I should say, if Desmond's in vitro technique had not been successful for *T. crassiceps* cysticerci.

For several years after Desmond's revealing discovery regarding "physiological" differences between horse and sheep hydatid cyst growth characteristics, the whole idea of "strain biology" took on a significant position in the parasitic helminth literature (note that the strain theory for other parasites/diseases, including *Plasmodium*/malaria, has been around for a much longer time; see McKenzie et al., 2008). The traditional lines of investigation that initially evolved regarding strain biology variously included "geographic distribution, host range, host specificity, metabolism, developmental rate, reproductive biology, growth in vitro, infectivity, morphology, and protein and isozyme analysis" or some combination (McManus, 1997). With the advent of modern molecular techniques and genetics, for example, polymerase chain reaction (PCR) and sequencing, substantial advances were made in our understanding of strain biology in this "cryptic" group and several other helminth species, including, especially those of the genus *Trichinella* (Posio, La Rosa, Rossi, and Murrell, 1992; Pozio, La Rosa, Murrell, and Lichtenfels, 1992).

Throughout the present chapter, the terminology has referred to "strains" rather than species. Identification at the strain level was first based on host specificity among intermediate hosts like that observed by Smyth and Davies (1974b) for *E. granulosus* in sheep and horses. Additional strains were subsequently added to the list, most of them involving *E. granulosus*, although the same kind of strain situation was also recognized for *Echinococcus multilocularis* but on a smaller scale. At one point in time, there were nine strains of *E. granulosus* identified variously with sheep, water buffalo, Tasmanian sheep, cattle, camels, pigs, cervids, lions, and horses. There were also four distinct morphospecies of *Echinococcus*, that is, *E. granulosus*, *E. multilocularis*, *E. oligarthrus*, and *E. vogeli*. The first two of these four species have cosmopolitan distributions, while the latter two are neotropical, restricted to Central and South America. Moreover, *E. vogeli* is polycystic, resembling *E. multilocularis* in its exogenous budding characteristics, while *E. oligarthrus* is unicystic and more like *E. granulosus*. For *E. vogeli*, the natural intermediate and definitive hosts are pacas and bush dogs, respectively. For *E. oligarthrus*, the intermediate hosts include opossums, agoutis, spiny rats, pacas, and rabbits, whereas definitive hosts are cougars and small felid species.

However, not long ago, Nakao et al. (2007) broke from the strain concept and stated that "all strains or genotypes of *E. granulosus* could [can] not be assembled into a monophyletic group, indicating that *E. granulosus* sensu lato is a cryptic species complex." Further, they continued, "The taxonomy of the cryptic species should be based entirely on their phylogenetic relationships, and appropriate scientific names should be used instead of strains and genotypes." Accordingly, they first recognized 10 genotypes (G1 to G10) within *E. granulosus*. They disengaged from the traditional strain model and identified the 10 genotypes as separate species, although they were not totally successful in their effort. They thus were unable to assign a species name to G6, G7, or G8; G9 is recognized as *E. multilocularis* and G10 as *Echinococcus shiquicus*, a relatively new polycystic species (see Xiao et al., 2002) that uses the Tibetan fox as a definitive host and plateau pikas as intermediate hosts.

I look back on my experiences with *T. multiceps* and *T. crassiceps* and wish in some ways that I had stayed with that line of work. I do not claim that I contributed anything to current phylogenetic thinking during my in vitro culture days working with Desmond Smyth, but I was there, and I watched while he and Zena Davies failed in their efforts to culture horse protoscolices (somewhat confirming my notion of in vitro culture as alchemy). However, I was particularly happy when she and Desmond figured out what had happened and provided an early hypothesis regarding the idea of "physiological strains" as applied to *E. granulosus*. That, folks, was really good science!

At this point, I break from my in vitro culture tale and describe my encounters and interactions with an English parasite ecologist by the name of Clive Kennedy. Before the London experience, I had known about Clive because of some excellent work he had done in a small lake called Slapton Ley, down near the coast of Devon. I first met Clive in Loughborough, the meeting site in the spring of 1972 of the British Society for Parasitology. I had, by that time, already spent several summers teaching at the W.K. Kellogg Biological Station in Michigan. I probably had not reached a decision yet, but my research had already produced a couple of papers dealing with parasite ecology. I recall spending time with Clive at the meeting. He actively encouraged me to make the switch out of my biochemistry/physiology line into parasite ecology. It was almost like he was "recruiting" me to make the break. Later in the spring, I went by train down to Exeter and spent an entire day with him. Except for my success with the in vitro culture study and the early *T. multiceps* taxonomic and experimental infection work, I had begun to realize that most of my research was more or less the same as others had done 5 years before me. In other words, I was not really on the cutting edge of anything.

I was beginning to feel that Clive was correct, because he had been telling me that ecological parasitology was almost a virgin territory (not including epidemiology, which had a huge head start on what I term parasite ecology). There were some parasitologists who had made great inroads in the area by then, for example, John Holmes, Jimmy Chubb, Harry Crofton, V.A. Dogiel, Gerry Schad, Wincenty Wisniewski, and, of course, Clive himself. What I did not know about was the truly significant impact that Harry Crofton's pair of 1971 papers was about to have on the whole field of parasite ecology and how the discipline was going to unalterably change over the next 10 years. His ideas regarding the concept of overdispersion and integration of mathematical modeling were to "set the stage" for the development of quantitative parasite population biology and ecology. Roughly 10 years later, the work of Bush and Holmes (1983, 1986a, 1986b) with helminths in lesser scaup ducks was to take our discipline into the complexities of parasite community ecology.

A lot of people are challenged to make a switch in their careers, for one reason or another. Some will do it, and others are too afraid. I too had to make this decision. By the time we left London, I knew what I must do. When we returned home, I had three PhD students waiting for me, that is, Joe Bourque, Herman Eure, and Bob

Morcock. The latter was almost finished with his dissertation work. For the other two, Joe and Herman, I gave an option, that is, follow the ecology route or go elsewhere. They chose to stay with me in ecology, but they took their research to the Savannah River Ecology Laboratory where Joe worked on turtles and Herman worked with largemouth bass.

All is well that ends well. I would like to also say that all is well that starts well or at least I had high hopes that it would! I can now say, without any doubt or question after more than 45 years of consistent work in ecological parasitology, I made the right choice.

References

Bush, A.O., and J.C. Holmes. 1983. Niche separation and the broken-stick model: Use with multiple assemblages. *The American Naturalist* **122**: 849–855.

Bush, A.O., and J.C. Holmes. 1986a. Intestinal helminthes of lesser scaup ducks: Patterns of association. *Canadian Journal of Zoology* **64**: 132–141.

Bush, A.O., and J.C. Holmes. 1986b. Intestinal helminthes of lesser scaup ducks: An interactive community. *Canadian Journal of Zoology* **64**: 142–152.

Esch, G.W., and J.D. Smyth. 1975. Studies on the in vitro culture of *Taenia crassiceps*. *International Journal for Parasitology* **6**: 143–149.

McManus, D.P. 1997. Molecular genetic variation in *Echinococcus* and *Taenia*: An update. *Southeast Asian Journal of Tropical Medicine and Public Health* **28** (Suppl. 1): 110–116.

McKenzie, F.E., W.P. O'Meara, D.L. Smith, and E.M. Riley. 2008. Strain theory of malaria: The first 50 years. *Advances in Parasitology* **66**: 1–46.

Nakao, M., D.P. McManus, P.M. Schantz, P.S. Craig, and A. Ito. 2007. A molecular phylogeny of the genus *Echinococcus* inferred from complete mitochondrial genomes. *Parasitology* **134**: 713–722.

Posio, E., G. La Rosa, P. Rossi, and D. Murrell. 1992. Biological characterization of *Trichinella* isolates from various host species and geographical regions. *Journal of Parasitology* **78**: 647–653.

Pozio, E., G. La Rosa, D. Murrell, and R. Lichtenfels. 1992. Taxonomic revision of the genus *Trichinella*. *Journal of Parasitology* **78**: 654–659.

Smyth, J.D., and Z. Davies. 1974a. Occurrence of physiological strains of *Echinococcus granulosus* demonstrated by in vitro culture of protoscolices from sheep and horse hydatid cysts. *International Journal for Parasitology* **4**: 443–445.

Smyth, J.D., and Z. Davies. 1974b. In vitro culture of the strobilar stage of *Echinococcus granulosus* (sheep strain): A review of basic problems and results. *International Journal for Parasitology* **4**: 631–644.

Xiao, N., J. Qui, M. Nakao, T. Li, W. Yang, X. Chen, P.M. Schantz, P.S. Craig, and A. Ito. 2002. *Echinococcus shiquicus* n.sp., a taeniid cestode from Tibetan fox and plateau pika in China. *International Journal for Parasitology* **35**: 693–671.

3 Gull Lake and the W.K. Kellogg Biological Station

The sun is coming down to earth, and the fields and waters shout to him in golden shouts.

The Ordeal of Richard Feverel, George Meredith (1828–1913)

Aside from John Larsh, who was head, Jim Hendricks was the senior faculty member in the Department of Parasitology when I arrived in Chapel Hill to begin my NIH Training Grant program in the fall of 1963. Jim's office was next door to mine on the floor that the parasitology department occupied in the School of Public Health. Knowing that I was still "wet behind my ears," he sort of took me under his wing, and I was greatly appreciative for all of his help. One of the things I noted very early was that while he was quite knowledgeable about medical parasitology, he was very much more oriented toward field parasitology than to the immunology being done in the department by his colleagues. He was not an ecologist, but was more of a natural history person, in the classic mold of people like Horace Stunkard, George LaRue, Lyle Thomas, Wendell Krull, and W.W. Cort (Jim Hendricks walked in some great shoes and did pretty well at it). All of these early parasitologists knew huge amounts about the parasites on which they worked—not just morphology, but their systematics, life cycles, and biology as well.

Jim taught field parasitology each summer at the University of Michigan Biological Station situated on Douglas Lake in the northern part of the Lower Peninsula. The field station was a magnet for many of the great parasitologists of the 1920s through the 1950s. Most of the second-/third-generation parasitologists (one to two

Ecological Parasitology: Reflections on 50 Years of Research in Aquatic Ecosystems, First Edition. Gerald W. Esch.
© 2016 John Wiley & Sons, Ltd. Published 2016 by John Wiley & Sons, Ltd.

generations behind H.B. Ward) would bring students with them to the station where they could be "traded," like chattel, or recruited by others for advanced degrees.

Among these early parasitologists was Don Wooten who brought his young students with him all the way from Chico State University in northern California to Douglas Lake every summer. Jim had recruited three of Wooten's students, Bruce Lang, Darwin Murrell, and Larry Gleason, into the Chapel Hill parasitology program. I was lucky to overlap all three of them, and because all four of us were still "rookies," we became great friends during my 2-year stint at the University of North Carolina.

In the spring of 1965, while still a postdoc and doing my work on *Taenia multiceps*, Jim came into my office one day and asked if I would like a job for the coming summer. Without thinking I responded in the affirmative, but immediately asked, "What kind of job?" He told me about a friend of his that he knew from teaching at the University of Michigan Biological Station. His name was George Lauff and he had just become the director of the W.K. Kellogg Biological Station (KBS) (of Michigan State University) on Gull Lake, about halfway between the cities of Kalamazoo and Battle Creek, in southwest Lower Michigan. George was interested in hiring someone to come in and teach field parasitology the following summer. He had already offered the job to Martin Ulmer (to handle the helminthology) and Norman Levine (to teach the protozoology). Jim said he had argued vehemently with George against offering the job to what he called a couple of "old prima donnas." Jim knew that I had experience in field parasitology having spent a summer at the University of Oklahoma Biological Station at Lake Texoma early in my graduate school days. He argued that George should give the job to a "youngster" instead of the two "old" men. George called me a few days later and told me that both Ulmer and Levine had turned him down and if I would be interested. "Yes," I responded immediately to the invitation. Before going any further, I must say that I cannot ever thank George enough for offering the best break in my early career. About 5 years later, I made a 180° switch in my research direction, in large part due to being hired by George to teach parasitology at KBS. I should also say that the financial aspect of the job could not have come at a better time. Support for my National Institutes of Health (NIH) Training Grant would end on 30 June. My first paycheck from Wake Forest would not come until 30 September—and Ann and I certainly did not have very "deep pockets."

Gull Lake is an extraordinarily beautiful body of water, about 6 miles long and a mile wide and roughly a 100 ft down at its maximum depth (Figure 3.1). The focal building at the station is W.K. Kellogg's (yes, Kellogg of Battle Creek) summer home— The Manor House (I was actually assigned to teach my parasitology course in it for a couple of summers). It is a very large, beautiful, two-story Tudor style home, situated on a high bluff on the east side of the lake. At that time, there were several other buildings at the station, for example, dormitories, an auditorium/cafeteria, a teaching/research building, a large boathouse, a carriage house, and so on (I am sure there are more now, but we have not been back since the early 1980s). The grounds were always well kept, and there was a small beach down below The Manor House

Figure 3.1   Gull lake adjacent to the W.K. Kellogg Biological Station in Michigan. Researchers at the lab had primary access to the boat present in the foreground.

for the exclusive use of anyone staying at the station. Families were housed in a two-story building in which were situated a dozen, or so, one-bedroom apartments, each with a bath and a small dining/living area adjacent to a kitchenette. Faculty did not eat at the cafeteria in those days, although if Ann had her way, our entire family would have taken all of our meals there. It was tight, with three children and the two of us, but we managed just fine. Our kids slept in the bedroom on bunk beds, and Ann and I used couches that doubled as beds at night out in the "living/dining" area.

The faculty was superb. George had brought together some of the best aquatic people in the country, including Don Hall, Don McNaught, Earl and Pat Werner, Bob Wetzel, Mike Klug, and Ken Cummins, to name just a few. Most of these guys were permanent faculty, with joint appointments in appropriate departments on campus and the station itself. Several others from the Michigan State campus, and elsewhere (including me), joined them in the summer. The offerings in various areas of field biology were comprehensive. I heard it said by a number of outsiders that KBS was considered as one of the best (if not the best) freshwater biological stations in the country (even when compared with the University of Michigan field station up at Douglas Lake).

Since I was to teach parasitology at a field station, I felt that I should emphasize field parasitology. During those years, there were no Animal Care and Use Committees, so this gave us so many more opportunities to collect all sorts of animals for laboratory necropsy and specimen preparation. While I had not yet switched into ecology, I nonetheless felt an obligation to emphasize the topic in a way that would embellish my parasitology focus. In fact, one of the main problems I had in those first years was that I did not know that much about any kind of ecology—remember my graduate student training had been in biochemistry/physiology.

The first parasite ecology book that I purchased was by a Russian, V.A. Dogiel, who had written a book titled *General Parasitology*. In the area of parasite ecology, the English, Russians, and Polish were way out in front of most others in the world in the mid-1950s and early 1960s. As I gained knowledge, however, and was beginning to think more like an ecologist, I would come to consider them more in terms of natural history than natural science.

Two different parasitologists published what I would consider the true "break-through" papers of modern parasite ecology. The first papers were by John Holmes in 1961 and 1962 and were focused on competitive exclusion. The really quantitative aspects of modern ecological parasitology were to appear when Harry Crofton's modeling papers were published in 1971 (Crofton, 1971a, 1971b). Although some modeling efforts were published prior to those of Crofton, I believe that most folks working in this area would agree that his were seminal.

There were several research studies with which I was directly connected during my 10-year tenure at Gull Lake. One of them lasted longer than 10 years, but the first was a turtle paper that examined the intestinal content for parasitic helminths from Whit Gibbons' painted turtles in Sheriff's Marsh. The second had to do with a comparison of parasite communities in an oligotrophic lake with those in a nearby eutrophic system. The third was a team effort involving several of my parasitology graduate students over a period that lasted for 20 years.

As is characteristic for field stations and biology departments, regular seminars are frequent and usually of real interest. This was certainly true for KBS. Sometimes, the speaker would be brought in from the outside and other times from within. My entry into ecological parasitology from a research perspective was in part also seeded by Don Hall by a seminar he gave in which he announced that Gull Lake was undergoing eutrophication and that something had to be done quickly.

I had just a vague idea of what eutrophication was at the time and what it had to do with parasitism. My earliest familiarity with the terms "eutrophic" and "eutrophication" came from a Polish parasitologist by the name of Wincenty Wisniewski who had conducted a series of studies on the parasite fauna of eutrophic Lake Druzno in Poland back in the 1950s (Wisniewski, 1958; Esch, 2004). Reference to the site of the study and to his fieldwork came directly from Dogiel's book that I cited earlier.

So, what is eutrophication and how does it relate to the ecology of host–parasite interactions? If one had taken a boat out to the middle Gull Lake in July during the very early 1960s, just before the beginning of the eutrophication process, and measured the dissolved oxygen (DO) at the surface and then at one meter intervals down to the bottom, a characteristic DO profile would have emerged. Beginning at the surface, the DO concentration would have been approximately 8 mg/L. It would have stayed that way down to about 10 m, then rapidly increased to approximately 12 mg/L, and remained that way down to the bottom. The temperature profile would have started at the surface at about 22–24°C down to about 10 m; then, within the next 2 m (and parallel with the change in DO), it would have sharply declined to about 6–8°C and continued that way down to the bottom. The explanation for the temperature shift has to do with the fact that as water becomes colder, it also becomes more dense, at least down to 3.96°C, at which point it is always most dense. The sharp decline in water temperature is reflected by a very strong temperature/density gradient (the metalimnion = thermocline) that effectively creates two lakes, the hypolimnion (below) and the epilimnion (above), with a metalimnion (thermocline) separating the other two from each other. In fact, so strong is the density gradient in the thermocline that it does not allow water in the epilimnion and hypolimnion to mix during the time of stratification each year. The greater quantities of DO in the hypolimnion are based on two gas laws, that is, colder water is able to hold more gas (oxygen), and the greater the pressure, the more gas (oxygen) water is capable of holding.

Oligotrophic lakes possess a suite of predictable characteristics, just like eutrophic lakes. For example, if phytoplankton samples had been taken from the surface of oligotrophic Gull Lake in the early 1960s, the species number would have been relatively high in comparison with eutrophic systems. However, the population densities of species present would have been low. If the dissolved phosphorous concentration in the water had been measured as well, it also would be somewhat small, which would account for the low densities in phytoplankton population sizes. Phosphorus (a plant nutrient) is the primary limiting factor in controlling growth in the phytoplankton community of aquatic systems.

Near Gull Lake is another impoundment known as Wintergreen Lake. It is the centerpiece for a nature reserve and is eutrophic, although the late Bob Wetzel's (2001) book, *Limnology*, describes it as hypereutrophic. Part of the reason for expanding eutrophic to hypereutrophic is related to the excessive fecal deposition (and phosphorous) by a resident flock of Canada geese, plus the fact that Wintergreen serves as a very good "stopover" place for migratory birds as they make their way north in the spring and south in the fall. The lake is also relatively shallow, with a large littoral zone occupied at the surface by extensive emergent vegetation.

To promote eutrophication in a body of water like Gull Lake, all one needs to do is add phosphorus. The source of the nutrient can be allochthonous (from outside, such as runoff carrying fertilizers used for nearby farming operations) or autochthonous (from faulty septic systems or the overuse of lawn fertilizers adjacent to the lake). In the case of Gull Lake, it was thought to involve both of the latter. What are the effects of adding phosphorus to the lake? The addition of nutrients simulates growth of phytoplankton populations, especially those species that favor eutrophic conditions. Gradually, over time, the eutrophic species will dominate the phytoplankton community. Most phytoplankton are short-lived, persisting no more than a few days. When they die, they sink to the bottom of the lake where they decompose, a process that requires oxygen. With the passage of time, the DO within the hypolimnion will decline and not be replenished until the next fall turnover, or mixis, of the lake. Recall that the hypolimnion and epilimnion are separated by the metalimnion and that oxygen in the top part of the lake will not be transferred to the bottom until mixis. During mixis in the fall, oxygen-depleted water in the hypolimnion will be merged annually with surface water of the epilimnion and oxygen will be reloaded throughout. When fall mixis occurs, the temperature also becomes isothermal from top to bottom. During winter in the very cold north, ice will form at the top and the lake will again become stratified. Water temperatures below the ice cover will rapidly increase from 0 to 3.96°C and then stay isothermal all the way to the bottom. Ice off in the spring represents the second mixis of these lakes. A lake that stratifies twice (winter and summer) is said to be dimictic. Most of the bodies of water south of Gull Lake will stratify just once, in the fall, and are called monomictic.

Based on this description, it is clear that the most significant natural abiotic factor affecting the biology of any body of water is the presence or absence of DO in the hypolimnion. If oxygen is present, then the lake will support a diverse benthic (bottom-dwelling) community of arthropods, molluscs, annelids, protozoans, and so on, plus a cold-water fishery (primarily salmonids and coregonids in the north) in the hypolimnion, extending all the way up to the thermocline. If nutrients are added to the lake/pond, however, anoxia will develop beginning at the very bottom. Why? The dead phytoplankton will decompose and decomposition is an aerobic process, which relies on oxygen in the hypolimnion. Excess oxygen depletion will drive the hypolimnion toward anoxia. When anoxia begins, organisms present in the benthic community or in the water column are faced with three options. They might be able to adapt through evolutionary transition, but this takes a long time. Alternatively, they may die and become locally extinct. Or, they may move, although some species are unable to follow the latter course of action. Those that are capable of moving will be subjected to elevated temperatures in the epilimnion and will probably die off due to physiological incompatibility or competitive disadvantage with species that are better adapted to the warmer temperatures.

It is no wonder then that Don Hall raised a really serious concern about the eutrophication process that had began in the mid-1960s in Gull Lake. It had to be stopped, even reversed, if much of the lake's oligotrophic flora and fauna were to survive. Fortunately, reversal of eutrophication can be accomplished if nutrient loading can also be reversed.

Wincenty Wisniewski's (1958) contribution to parasite ecology was huge because it was one of the first steps in the evolution of ideas that led into the development of landscape epidemiology/epizootiology. As stated by Esch et al. (2008),

> The whole idea of landscape ecology (epidemiology/epizootiology) rests upon the non-random distribution of parasites in spatial terms, and of helminths in terms of host populations. In other words, parasites are generally present in certain areas of lakes or ponds (and terrestrial habitats as well) for a reason. For the most part, it is because hosts necessary to complete their life cycles are there, either as permanent residents or as ephemeral, but regular, visitors.

Wisniewski and his students had conducted a massive survey of the parasite fauna in eutrophic Lake Druzno in northern Poland. Their basic question was, does eutrophication affect the parasite fauna in this lake? Their work revealed the dominant presence of 67 cestode and 84 trematode species. In birds alone, they recovered 41,453 cestodes and 10,664 trematodes. The next most abundant helminths were acanthocephalans, with just 330 specimens. The least abundant were nematodes, with only 17 individuals.

Based on these observations, Wisniewski (1958) drew three conclusions that would apply to any body of water, plus one that focused only on eutrophic habitats. First, he said that the "… final hosts of tapeworms, flukes and thorny-headed worms [acanthocephalans] are a sort of concentrating sieve in a water biocoenosis, while intermediate hosts serve mainly to help these parasites pass to their final hosts proper." Second, he noted the heterogeneous distribution of parasites and their hosts within the lake. As he put it, "They occur in a greater congestion in some points." The third generalization indicated that some of the parasites were typical of eutrophic systems and some were not. These three observations cannot be challenged because they have been confirmed many times since his publication in 1958. The fourth assertion was that "In eutrophic bodies of water, particularly shallow ones, the parasitofauna of birds prevails and is characteristic of them." This statement is true, but the cause and effect implied for eutrophication and the uniqueness of the parasite fauna are not. In contrast to Wisniewski, Esch (1971) and Esch et al. (2008) declared that "While eutrophication may influence trophic dynamics associated with helminth transmission, it is the extent of the littoral zone (and shallowness of the lake or pond) that supercedes, and ultimately influences, the nature of the parasite fauna."

There is no question that during the late 1960s and early 1970s, Wintergreen Lake was eutrophic, even hypereutrophic, as asserted by Wetzel (1975). Likewise, although Gull Lake was undergoing eutrophication during the same time period, it had clearly not reached a eutrophic state, evidenced by the fact that it retained many of the same overall qualities of its original oligotrophic status. Esch (1971) proposed that the diversity of the parasite fauna in a given body of water is largely determined not by the eutrophic/oligotrophic state of the lake or pond, but much more likely by the nature of the trophic dynamics involving the dominant predators. In Gull Lake, the dominant predators at the time were smallmouth bass, pike, gar, and trout. The parasitofauna in these species is dominated by parasites completing their life cycles in these hosts. The littoral zone in Gull Lake is also limited, which is not favorable to most fish-eating birds in this kind of aquatic ecosystem. Comparatively, it is a "closed" ecosystem. In Wintergreen Lake, the parasite fauna is dominated by species that complete their life cycles in fish-eating birds and other waterfowl. In contrast to Gull Lake, Wintergreen Lake is an "open" ecosystem. Lake Druzno in Poland was eutrophic and dominated by parasites that completed their life cycles in fish-eating birds, but not because it was eutrophic. The Polish lake was, simply put, shallow and favored this kind of trophic interaction, just like Wintergreen Lake.

The observations described for Gull and Wintergreen Lakes were based on data collected from 1965 to 1970 and reflect activity at the ecosystem level. By accident, however, we also became involved with the eutrophication situation in Gull Lake while working with a single host–parasite system. During the late 1960s, a few of my Wake Forest graduate students accompanied me to KBS where they served as teaching assistants (TAs) in my parasitology course. In 1969, my TA was Bob Morcock, who was also working on his PhD degree with me at the time. During that summer, Bob, on an impulse, collected some subimagoes of the burrowing mayfly, *Hexagenia limbata* (Ephemeroptera). This insect was also parasitized by metacercariae of the allocreadiid trematode, *Crepidostomum cooperi*. In order to understand the eutrophication/parasite interaction involving *H. limbata* and *C. cooperi*, it is first necessary to describe the life cycles of the insect and the trematode.

In Gull Lake, the life cycle of this mayfly takes about 1 year to complete. At their beginning, fertilized eggs are deposited on the lake surface by nonfeeding adult females, after which the latter immediately die. Quickly, eggs hatch into juveniles that construct U-shaped tubes in the lake's substratum. The external gills on their abdomen wave to and fro inside the tubes and create a current, into which detritus is pulled from the benthic substratum. It is from this detrital material that the subimagoes obtain their sustenance. Throughout the next 12 months (at least in Gull Lake—further north, the cycle takes 2 years to complete), the insects grow and molt something on the order of 28–29 times all the while moving deeper and deeper into the lake. Just before the next to last molt, the juveniles exit their burrows, swim to the surface, and change into winged subimagoes (sexually immature adults), which fly off into vegetation along the edge of the lake. Within another 24 hours, they will molt one more time to the adult stage. The evening after the final molt, males form a mating swarm along the lake's shoreline through which females fly. As they pass by, males catch and inseminate the females while still on the fly and then release them, permitting the impregnated female to fly out onto the lake's surface. There, females oviposit and die, while males land on vegetation and die the same evening. During oviposition, the females are also totally vulnerable to predation by fish. In fact, the insects apparently create enough disturbances on the surface water while ovipositing that they attract piscine predators. The stomachs of several fish species in the lake will be full of mayflies the next morning. During the 10 summers I taught at the lab, mayfly emergence always began within a few days after 1 August, peaked quickly, then declined over the next few weeks.

Crepidostomum cooperi is a common digenetic trematode among centrarchid fishes in Gull Lake, where, as adults, they occupy the pyloric caecae. When eggs are shed, they pass to the outside in feces where they hatch after several days of embryonation. Free-swimming miracidia then emerge from the egg and enter fingernail clams through their incurrent siphons. After developing into sporocysts and then rediae (both juvenile stages), opthalmoxiphidioleptocercous cercariae (another juvenile stage) are produced by rediae, with cercariae being released from the clam. The extended terminology associated with naming the cercaria type simply means that *C. cooperi* cercariae are equipped with eyespots, a stylet, and a thin tail. What this also tells you is that the cercaria is sensitive to light, it possesses a tool to assist in penetration of tissue, and it is capable of swimming. Experimentation has demonstrated that the *C. cooperi* cercariae are negatively phototactic (and probably positively geotactic). In other words, they swim close to the lake substratum. When they swim over one end of a mayfly tube, the cercariae will be sucked in and pulled to the level of the external gills. Here, they attach and penetrate the gill surface. After shedding its tail, the cercaria will secrete a cyst wall and attract melanocytes that cause the cyst wall to turn black. Because the metacercariae are dark in color and occur just below the cuticle of the subimago and nymph, they are simple to count.

Subimagoes are easily collected during the evening and night of their emergence from the lake (they are attracted to light) and placed in 70% ethanol for storage. Leaving a light on in a lab at night will insure that large numbers of subimagoes will accumulate on window screens. When we began collecting mayflies, we simply picked them off the screens with our fingers and placed them in alcohol. Later, when we sought to increase the size of our collections, we used a vacuum sweeper. It was easy "pickins!"

When we returned to Winston–Salem and Wake Forest in the late summer of 1969, Bob gave me the mayflies so I could count the metacercariae, which I did, even though I had no real reason for doing so, at least not initially. This gave us the first data points for a study that was to last for 19 more years. It is worth noting that initially I had no intention of extending the work. At that point in time, however, I was just beginning to understand the nature of the eutrophication process. Moreover, I also knew very little about the mayfly and its biology. For some reason, however, mayfly subimagoes were collected the next summer and then again. I am really not certain when the connection between eutrophication and biology of the mayfly and trematode was made, but it was, eventually, and it made for an interesting story.

Now then, what makes the *C. cooperi/H. limbata*/eutrophication combination so appealing that we were willing to spend 20 years pursuing it? The answer rests entirely with the biology of the host and parasite and its association with the eutrophication process in the lake. Before getting to the data, let me briefly identify my "long-term" collaborators, four graduate students, that is, Terry Hazen, Amy Crews, Tim Goater, and Dave Marcogliese. At the time, Terry was working on the biology of *C. cooperi* in amphipods as part of a master's degree project at Michigan State University. He subsequently began working on his doctoral dissertation and red sore disease at the Savannah River Ecology Laboratory in South Carolina during that summer of 1975, completing it 3 years later at Wake Forest. As an undergraduate at Wake Forest, Amy took my parasitology course and decided that she wanted to do graduate work in my lab, which she did for her MS degree. I will talk more about her later too because she was the first of my graduate students to work in a place called Charlie's Pond. Tim Goater came to Wake Forest University from Al Bush and Brandon University in Canada to pursue an MS degree, which he completed in 1985, and then a PhD degree, which he finished in 1989, also in my lab. Dave Marcogliese was another Canadian who came down from Montreal, via Halifax, Nova Scotia, to do his PhD at Wake Forest.

As I said earlier, work on the mayfly/*C. cooperi* system in Gull Lake was begun in 1969 with the mayflies brought back to Winston–Salem by Bob Morcock. Mayflies were collected during their emergence from the lake in August of each summer from then through 1988. Terry and I had undertaken some seasonal work on *C. cooperi* in both nymphs and subimago mayflies in 1981 (Esch and Hazen, 1982). Our results from the mayfly and amphipod work indicated that *C. cooperi* recruitment begins in June and continues through August and then stops. The parasite disappears from centrarchid definitive hosts in late fall and, most likely, overwinters in the fingernail clam first intermediate host.

During the spring of 1984, Amy, Tim, Dave, and I began talking about the mayfly/*C. cooperi* system and the eutrophication situation in Gull Lake. It was at this point I invited them to become part of the project (I have always believed that it is more fun if you have collaborators). We knew that a sewer system surrounding the lake was operational, having been constructed in 1983. We also knew that Bob Wetzel (pers. comm.) had predicted recovery in the lake would take about 5 years to complete. In other words, if we wanted to acquire data regarding population biology in mayfly nymphs and subimagoes during the time of peak eutrophication, our "window of opportunity" was going to close soon, within just a few years.

Accordingly, all four of us traveled to Michigan early in the summer of 1984. Since mayfly subimagoes were not available until later in summer, our focus was on mayfly nymphs. Over a 5-day period, we sampled pairs of sites at depths of 3, 5, and 7 m. As we had predicted, the mean prevalence and density of *C. cooperi* were lowest at

7 m and highest at 3 m. Moreover, while there was some variability in both prevalence and density of the parasite between 1969 and 1984, the trend in both parameters was clearly upward. In fact, the highest prevalence occurred in females from 1984, while it was second highest in males. In 1985, there were very slight changes in the same two parameters among both males and females. However, the next 3 years tell an entirely different story but a confirming one with respect to our predictions regarding the parasite population biology in concert with eutrophication reversal in the lake. In the 3 years following 1985, prevalence in males declined from 90.6% (1985) to 48.0%, 34.4%, and 25.0%, respectively; in females, prevalence dropped from 99.0% to 79.4%, 52.6%, and 37.6%, respectively. Metacercariae densities in males dropped from 5.6/host to 1.5, 0.9, and 0.4, respectively. In females, densities declined from a peak of 16.5/host to 3.4, 2.9, and 1.0, respectively, in the next 3 years (Esch et al., 1986; Marcogliese et al., 1990). Bob Wetzel was definitely on point regarding the reversal and timing of recovery in our lake. We were also correct regarding the way in which change should occur in *C. cooperi* population biology.

So, our research in Gull Lake began with a few mayflies being haphazardly collected one summer night in August of 1969 by one of my graduate students (Bob Morcock) and ended 20 years later in August of 1988, with the collaboration of four graduate students (Terry Hazen, Tim Goater, Dave Marcogliese, and Amy Crews). Although the previous story is an interesting one (to me, at least), the exciting part was my opportunity to include three of them in what I call "gravy" research, that is, something that comes on top of their MS/PhD work and any publications that these efforts should produce. I have always considered the gravy papers to be the best kind because they are not planned—the research is accomplished and publications emerge because someone sees an opportunity and jumps on it!

References

Crofton, H.D. 1971a. A quantitative approach to parasitism. *Parasitology* **62**: 179–193.

Crofton, H.D. 1971b. A model for host-parasite relationships. *Parasitology* **63**: 343–364.

Esch, G.W. 1971. Impact of ecological succession on the parasite fauna in centrarchids from oligotrophic and eutrophic ecosystems. *American Midland Naturalist* **86**: 160–168.

Esch, G.W. 2004. Parasites, People, and Places. Cambridge University Press, Cambridge, 235 p.

Esch, G.W., and T.C. Hazen. 1982. Studies on the seasonal dynamics of *Crepidostomum cooperi* in the burrowing mayfly, *Hexagenia limbata*. *Proceedings of the Helminthological Society of Washington* **49**: 1–7.

Esch, G.W., T.C. Hazen, D.J. Marcogliese, T.M. Goater, and A.E. Crews. 1986. A long-term study on the population biology of *Crepidostomum cooperi* (Trematoda: Allcreadiidae) in the burrowing mayfly, *Hexagenia limbata* (Ephemeroptera). *American Midland Naturalist* **116**: 304–314.

Esch, G.W., K. Luth, and M. Zimmermann. 2008. Wincenty Wisniewski: A retrospective view. *Wiadomosci Parazytologiczne* **54**: 279–281.

Marcogliese, D.J., T.M. Goater, and G.W. Esch. 1990. *Crepidostomum cooperi* (Allocreadiidae) in the burrowing mayfly, *Hexagenia limbata* (Ephemeroptera), related to the trophic status of a lake. *American Midland Naturalist* **124**: 309–317.

Wetzel, R.G. 1975. Limnology. W.B. Saunders Co., Philadelphia, PA, 743 p.

Wetzel, R.G. 2001. Limnology: Lake and River Ecosystems. Academic Press, San Diego, CA, 1006 p.

Wisniewski, W.L. 1958. Characterization of the parasitofauna of an eutrophic lake (parasitofauna of the biocoenosis of Druzno Lake, part I). *Acta Parasitologica Palonica* **6**: 1–64.

4 Gull Lake and the Connection with the Savannah River Ecology Laboratory

For my part, I travel not to go anywhere, but to go. I travel for travel's sake. The great affair is to move.

Travels with a Donkey, Robert Louis Stevenson (1850–1894)

Some of the research that begun in Gull Lake during the mid-1960s was soon to be linked to work at the Savannah River Ecology Laboratory (SREL) near Aiken, South Carolina. It was to involve several of my own students and also a graduate student at Michigan State University from Alabama by the name of J. Whitfield (Whit) Gibbons.

I had just arrived for the first summer of teaching at the Kellogg Biological Station (KBS) in 1965 when I met Whit. He was working toward his PhD on the population biology and ecology of painted turtles in Sheriff's Marsh close by the field station. One of the important parameters he had to measure was clutch size in female turtles. The only way he could do this was to kill the turtles, open them up, and count the eggs. Until I arrived, the turtle carcasses were disposed of as garbage. When I learned what he was doing with the digestive tracts, I asked if he would give them to me so I could check for parasites. I recall his reaction when I asked—it was like, "Can I trust you?" He eventually consented and was to become a coauthor on our paper dealing with the endoparasite fauna in painted turtles (Esch and Gibbons, 1967). Although we were able to define the parasite fauna fairly well, we were unable to establish any clear patterns of seasonal change because the sample sizes were just too small. Nonetheless, it was Whit's first scientific paper and my first "parasite ecology" paper (even before I made the switch from biochemistry/physiology to ecology).

Ecological Parasitology: Reflections on 50 Years of Research in Aquatic Ecosystems, First Edition. Gerald W. Esch.

Whit and I became very good friends and still are. The University of Georgia hired him when he completed his PhD degree. While he did become a member of the biology faculty in Athens, he was posted at the SREL, on the site of the Savannah River Plant (SRP) not far across the river from Augusta, Georgia. Gene Odom who was already a world-class ecology researcher started the ecology lab. In fact, his classic terrestrial succession studies were actually conducted at the Savannah River site (field 3412). Although Whit was not one of my graduate students, I think we hit it off well because we were so close in age and experience. When we met, I had just completed my PhD and a 2-year postdoc, and he was almost done with his PhD. He is retired now and I am close to the same status. If you "google" his name, you will find that he is one of the foremost herpetologists in North America and still active!

The SRP is situated on a parcel of land about 300 square miles in size, bordered on the south by a cypress swamp and the Savannah River and on the other three sides by a 10 ft high chain-link fence. The plant was the location of a heavy water manufacturing facility and five nuclear reactors for the production of weapons-grade plutonium. Well-armed guards manned all entrances to the site, making sure that anyone who entered possessed an appropriate identification badge. Each person with such a badge had been carefully screened and checked by the FBI.

At one place on the SRP, there had been a small town, named Ellenton. I am not sure why, but when the land was acquired by the government, every house was torn down or the houses were moved to a new site; garages, barns, schools, stores, and so on were also destroyed completely. The old town of Ellenton was renamed New Ellenton and is situated about 2 miles north of the plant and a few miles south of Aiken. An odd thing about the old Ellenton location is that, although all of the structures are gone, everything painted on the streets of the former town remains, and you can see occasional steps and chimney remains at some sites where houses had been located. You can also still see "railroad crossing," "slow, school crossing," "hospital" designations, and so on painted on the streets, which were still intact. The "spookiness" of the site is exacerbated by the fact that weapons-grade plutonium was being produced close by. Its production was a reminder that a town was once there and that the power of an atomic explosion would easily create nothing but painted reminders on the streets of what used to be a town called Ellenton.

Whit was to spend his entire career at SREL doing various sorts of herpetology research. In total, five of my graduate students, Herman Eure, Joe Bourque, Joe Camp, Kym Jacobson, and John Aho, successfully completed their PhD or master's fieldwork at SREL, with Whit as the local supervisor. Another student, Terry Hazen, and I spent a full year there in 1974–1975 working on red sore disease in alligators and largemouth bass. I will subsequently describe some of their/our research down at SREL, but before I do, I want to take you back to Gull Lake and another project because it is what initially tied SREL and Gull Lake together.

In 1967, Joe Johnson, a facilities technician at KBS, and I began to monitor the presence of larval and adult *Proteocephalus ambloplitis*, a tapeworm in smallmouth bass from Gull Lake; our data were generated from April to October in 1967, 1968, 1972, and 1973 (Esch et al., 1975). Jim Coggins had been sent to me as a graduate student by Jim McDaniel, one of my very close friends from Dr. Self's lab at the University of Oklahoma. Jim Coggins joined Joe and me during the last 2 years of collecting *P. ambloplitis* from bass. It was a great opportunity for Jim because it was to lead into his dissertation research, which was focused on light and transmission electron microscopy of *P. ambloplitis*, primarily the plerocercoid (juvenile) stage.

The life cycle of *P. ambloplitis* was initially described by Hunter (1928). He asserted that the adult tapeworms shed eggs into the water, where they were ingested by cyclopoid copepods. On hatching from the egg, a larva penetrated the gut wall, entered the hemocoel, and developed to the procercoid stage. The infected copepod was then ingested by a planktivorous fish where the procercoid penetrated the fish's gut wall and migrated into the visceral mass, where it developed to a plerocercoid. Smaller fish with plerocercoids were said to be ingested by larger ones, where the worms would develop into mature worms in the intestines of bass hosts, thus completing the life cycle. For 40 years, this life cycle description was widely accepted as accurate by parasitologists until Hertwig Fischer and Reino (Ray) Freeman (Fischer and Freeman, 1969) published a paper indicating that Hunter's description of the life cycle of *P. ambloplitis* was incorrect.

The work of Fischer and Freeman (1969) provided strong evidence from both fieldwork and laboratory experimentation for a phenomenon called "internal auto-infection." Their findings indicated that plerocercoids (at least some of the larvae present in the visceral mass) would migrate into the intestine from their parenteric sites when the bass reached greater than 10.7 cm in length and water temperatures increased from 4 to 7°C. They felt since total host length had to be greater than 10.7 cm and water temperature had to move from 4 to 7°C, "... that other factors such as the hormonal state of the host, for example, may enhance the effect of temperature" (Fishcer and Freeman, 1969). Of course, they were implying that the plerocercoid migration (=internal autoinfection) was influenced in some way by the spawning of bass. If this idea were extended further, it would suggest that migration could only occur in the spring and that plerocercoids acquired after spawning was completed would accumulate until the next spawning season. In other words, Fischer and Freeman had definitely shown that plerocercoid migration in adult bass was a form of internal autoinfection and that adults did not develop directly from plerocercoids directly acquired by ingestion of plerocercoid-infected fish (a trend observed for other species of *Proteocephalus*).

When I arrived at the KBS in 1965 to teach field parasitology, I decided that a requirement for students in the course would include the submission of their own parasite collection, consisting of a certain number of keyed helminths from different parasite groups. Part of my reason for making this requirement was to give students the experience of staining their own specimens, learning to use keys and the primary literature in making identifications, and so on. Vicariously, it would also give me a quick picture regarding the local parasite fauna. We were on a lake, so, naturally, fish parasites became a primary focus. In fact, almost any aquatic species became a target. Of course, this also meant setting gill nets the night before teaching and then running the nets early in the morning of class. Smallmouth bass were frequently caught and *P. ambloplitis* frequently turned up in our fish. As mentioned earlier, Joe Johnson and I began keeping careful track of adult *P. ambloplitis* obtained from smallmouth bass in 1967 and 1968, before the 1969 paper by Fisher and Freeman. In 1972 and 1973, after Fischer and Freeman's paper was published, we again kept accurate field records for *P. ambloplitis* in bass. During the first two collecting years, we thought that the life cycle of *P. ambloplitis* proposed by Hunter (1928) was accurate, but in the second 2 years, we knew that Fischer and Freeman (1969) were correct.

Our findings corroborated some of what Fischer and Freeman (1969) had seen, that is, the adult parasites disappear in the winter and are recruited in the spring, at least in the north. Our hypothesis for explaining this phenomenon followed that of Fischer and Freeman (1969) who speculated that spawning temperatures/spawning hormones were the triggers for recruitment of adult worms. Logically, this would make sense and would also explain the absence of adult parasites in sexually immature bass.

However, we all were wrong!

In the spring of 1969, I was appointed to chair our Department of Biology's Graduate Committee. I was sitting in my office after graduation when I heard a knock on my door. I looked up, and standing there was a handsome, young black man. He introduced himself as Herman Eure and told me that he would like to come to do graduate work in our department. Moreover, he wanted to do it in parasitology. Wake Forest University, like a number of other schools in the South during that period, was actively recruiting black students at the undergraduate level but very few into our graduate programs. However, by that time of the year, all of our TAs for the next year had been assigned, so I thought we had a problem. I explained to Herman, "We can assure you a scholarship, but our TA assignments have already been made for next fall." He looked at me sort of quizzically and said, "Thanks, but I don't need a TA. I have a Ford Foundation Fellowship that provides tuition, some money for supplies, some for travel, plus an unlimited stipend of $250/mo" (and remember, that was in 1969 dollars—not too shabby!).

Herman was formally admitted in the fall of 1969, and he has been one, or more, of the following ever since, that is, my student, friend, colleague, employee, boss, and friend. I left Wake Forest in 1974 for a leave of absence at the SREL, and Herman, with his PhD completed, was hired in my place and then kept on as a tenure-track assistant professor when I returned to the faculty in 1975. I refer to Herman as boss because he was to become chairman of our department several years later. I mention friend twice, not because he became a friend twice, but because he was to become a really close friend over the years.

When Herman became a graduate student in 1969, he decided to skip the master's degree and pursue the PhD directly. This meant that for the next couple of years, he would be focused on coursework. In the meantime, we had a number of discussions about his dissertation research. By the time I returned from London in the spring of 1972, I had made my decision to switch from biochemistry/physiology to ecology. I told Herman about changing course with respect to my new research focus, and he was excited about the prospect. At the same time we recruited Herman in 1969, Joe Bourque, another graduate student of mine, completed his master's degree. In 1970, he was admitted into our PhD program with the support of a Division of Nuclear Education and Training Research Stipend from the Atomic Energy Commission and in 1971 an Oak Ridge Predoctoral Fellowship for study at SREL. In the meantime, I had consulted with Whit Gibbons and asked if these two guys could do their research at SREL and if he would serve on their PhD committees—he agreed!

So, we were set to go. My only problem was that neither Herman nor Joe had ever been to SREL (neither had I). We decided that we should drive down and take a look at the place. We were scheduled to meet Whit at an entrance gate to the "bomb plant" (so-called by the locals) but arrived early for our meeting. So I suggested we stop and get a cup of coffee somewhere in downtown Aiken to kill some time. I was driving and pulled up and parked in front of a Woolworth's store. It was not until we got out of the car and started to go in that I thought about it. The sit-ins in the South had actually started in a Woolworth's store in Greensboro, North Carolina, in the early 1960s about 20 miles east of Winston–Salem and Wake Forest University, and here we were, two white guys and a black guy going into a Woolworth's in Aiken, South Carolina, a really southern town in what I considered to be the Deep South. Neither Herman nor I said a word. We walked in and sat down on stools at the breakfast counter and waited for the waitress. She came over to where we were sitting, took our order, and, within a few minutes, pleasantly served us. We drank our coffee, without incident, and headed out to the plant.

An interesting story involving Herman and the town of Aiken occurred prior to his move to Aiken following the completion of his course work at Wake Forest. Herman, Joe Bourque, and I, along with Whit, went to see the person who had agreed to rent an apartment to Herman and his family during their stay in Aiken. Her name was Shirley Pelham and she lived in a large house on Haynes Avenue, one of the main streets in the town. The place was classic antebellum, with Spanish moss on the trees and a garage apartment that she was going to rent to Herman. She informed us that she was originally from Pascagoula, Mississippi, but saw no problem renting to Herman and his family. As Herman explained to me, here was a situation where a white female landlord showed no particular concern that a black tenant would live in her garage apartment in the middle of the "white" section of town. Obviously, Herman continued, "… many of our preconceived notions about the South had to be altered."

Whit met us at one of the SRP gates and we were inside in no time. I must say that going into the field with Whit was quite an experience. By this point, he was settled in pretty well and wanted to take us to some of his research sites. One of the first places we visited was a "Carolina Bay." Scattered over North and South Carolina and Georgia are a substantial number of these shallow ponds, each one shaped like a tear drop with the large end always on the northwest and a small end pointing southeast. Some folks have asserted that they were created "eons ago" by a meteor shower of some sort. They are mostly ephemeral bodies of water that are filled with the soaking rains of spring and by fall and winter are mostly dry. While filled with water, they are inhabited by huge numbers of frogs, salamanders, and turtles. Whit had constructed a "drift fence" and a series of pit traps around one of these bays, about a mile in total distance. Every 5 m, or so, a five-gallon bucket was planted in the ground on either side of the drift fence. Anything that was moving in or out of the bay would eventually fall into a trap and be picked up the next morning, and the species would be recorded and then released on the other side. Literally thousands of amphibians and reptiles would be caught during each spring. Whit had also returned to his beloved turtles for extended research, this time yellow-bellied sliders. He would later resolve the problem of determining clutch size without killing the turtles by taking females to a lab where they would be x-rayed. Shadows of eggs were easily counted on the x-ray film. He also figured out how to identify individual turtles using a Xerox machine to record the rings and distinctive plastron features (much simpler than the old way of notching the plastron with a file).

Later that morning, Whit took us down to the cypress swamp contiguous with the Savannah River. We left our van and walked a few meters, and by then, Whit was literally into the swamp. Herman, Joe, and I were not dressed for fieldwork, so we carefully tiptoed out over the swamp on a fallen pine tree. Whit asked if we would like to see a water moccasin and of course, more or less in unison, we said yes. Pointing to the darkly stained swamp water, he said, "There's one, and there's one." Aiming a few meters away from the log, he exclaimed, "and, there's another." I still cannot figure how three grown men could move along a downed pine tree so fast that we almost ran to get back to dry ground.

We then were driven over to Par Pond, the largest of several reservoirs used as cooling systems for two of the reactors (Figure 4.1). These impoundments were necessary since water emerging from the reactors was in excess of 95°C and a lot of quick evaporation was taking place. Whit had brought a couple of fishing rods with us so he could show us how easy it was to catch largemouth bass in the reservoir. I have never seen, or heard of, anything like what we were to experience as we fished. It was absolutely incredible, nothing but bass greater than two pounds were reeled in, probably 20, in about 30 minutes. I recall another time when I stood on top of the dam at the north end of Par Pond and caught 40+ bass that were two to five pounds each, all within about 45 minutes. My arms were so tired and sore when I finished that I could barely raise them to the height of my shoulders.

Figure 4.1   Par Pond on the Savannah River Plant site in South Carolina. In the foreground is a "shocking boat." The craft was equipped with a generator for producing electricity used for stunning fish in water ahead of the boat. The electroshocked fish are stunned and rise to the surface where they are removed from the water with a dip net and then deposited in a large tub carried in the boat. In an hour of shocking, 50 largemouth bass could be easily captured, without killing them.

Our trip back to the lab turned into another "event." As we were driving along on the highway, Whit suddenly slammed on the brakes and told me to take the wheel. Before the van stopped, he was out the door and running across the road. He left the vehicle in gear and from my "shotgun" seat I had to hit the brakes and put it into park. We all turned to see what made him move so quickly and observed a coach whip snake moving quickly across the road toward the grass. We watched as Whit moved behind the snake, then reached down, grabbed its tail, and literally whipped the body of the snake back through his legs, which he quickly then closed together. He swiftly pulled the snake's body back through his legs and when the head appeared he grabbed it with his free hand. Whit then stood up, grinning like a "Cheshire cat," and walked back. When he reached the van, he asked if there were any volunteers to hold the snake as we drove back to the lab, but the response was a unanimous no! So, he grabbed an old wading hip boot on the floor of the truck and forced the snake into the boot. He handed me the boot and told me to squeeze the top of it tightly as we drove back. When we parked at the lab, Whit took the boot, turned it over, and shook it, but oddly, nothing came out. When he looked inside, there was no snake, but there was a hole about the size of a golf ball. All of us searched the truck high and low, but the critter could not be found. We never did figure out what became of the snake!

After the trip, both Joe and Herman decided that they would do their dissertation research at the SREL site, with Herman working on the population biology of helminth parasites in largemouth bass and Joe doing the same thing in yellow-bellied slider turtles, *Trachemys* (*Pseudemys*) *s. scripta*. The research of Joe and Herman developed into a partnership in many ways, mainly because they were able to help each other in the field.

Earlier, I referred to one of the cooling impoundments as Par Pond, so designated for reactors "P" and "R." Actually, it is not a pond at all. It is a very large reservoir, about 1100 hectares in surface area. Water from the reservoir is drawn into a reactor at one end and recycled back into the pond at the other end. Replacement water due to rapid evaporation came from the nearby Savannah River. The reactors produce water at very high temperatures (95°C), but there are also periods when the reactor power is diminished and temperatures decline. Overall, it is estimated that the average hypolimnetic temperatures in the heated part of the lake are 5°C above those of similar sized impoundments in the southeastern part of the United States. The maximum water temperatures in the heated end reached 45 and 35°C elsewhere during July.

As described earlier in this chapter, studies by Fischer and Freeman (1969) in Lake Opeongo (Canada) and Esch et al. (1975) in Gull Lake (Michigan) suggested the life cycle of *P. ambloplitis* in South Carolina should be seasonal, as it was in the north. If so, the pattern of appearance and disappearance of adult *P. ambloplitis* should at least parallel what we observed in the northern localities, that is, prevalent in late spring and summer before disappearing in early fall. However, this was not the case. In fact, it was almost the reverse (Eure, 1976). Adult *P. ambloplitis* were observed in sexually mature largemouth bass from unheated areas of Par Pond from December through May before disappearing in the summer. In the heated end of the reservoir, adult tapeworms were present from December through August before vanishing for 3 months. In other words, the life cycles, in terms of the presence of adult tapeworms, are almost exactly opposite in South Carolina and Michigan.

These data clearly suggest that spawning hormones are not involved in the recruitment of plerocercoids from parenteric sites into the intestine since largemouth bass spawn only in the spring at both northern and southern localities. On the contrary, there is somewhat of a parallel with rising temperatures. However, the temperatures at the heated end of Par Pond versus ambient localities in the lake did not really match each other over time. In fact, the mean temperature at the heated end ranged from 23 to 27°C from December through April, while ambient temperatures in other parts of the lake were approximately 13°C in December and January, and then they began climbing in February into the summer. In other words, there are great inconsistencies within heated areas and between heated and ambient localities throughout the year. Herman suggested (Eure, 1976) that perhaps a radical change in water temperature (up or down), as implied by Fischer and Freeman (from 4 to 7°C in Canada and Gull Lake in Michigan), stimulated the move from parenteric to enteric sites. However, to get anywhere near these temperatures, there would have to be a decline in South Carolina water temperatures and an increase in water temperatures in the north (Canada and Gull Lake), causing migration to enteric sites in the winter months in South Carolina and the spring months in Canada and Michigan.

So, what hypotheses can be offered to explain the life cycle discrepancies? Ever since the work in Gull Lake (Esch et al., 1975) and Herman's dissertation (Eure, 1974), I have wondered about resolving the question. However, the opportunity of conducting meaningful research never presented itself until Kyle Luth was ready to pursue his PhD. Kyle accepted the challenge and is on the track for resolving the issue. He and I have had a number of very long discussions regarding possible explanations for the differences between north and south. Morphologically, the cestodes in the two lakes are indistinguishable. We believe, therefore, that we might be dealing with cryptic *P. ambloplitis* species in the north and the south. So, how does Kyle test the hypothesis? The answer, we hope, is very simple. He has since obtained specimens from the north and south, with which he will use PCR to amplify three to four genetic loci under varying levels of selective pressure and then sequence the amplified DNA to compare these loci between fish from different lakes/ponds. He has added some additional goals for his dissertation work by asking how much genetic variability is there in various bodies of water, if genetic drift occurs in smaller farm ponds, and so on.

Toward this end, during the summer of 2012, he and Mike Zimmermann (another of my PhD students) sampled centrarchids in various ponds and lakes in 11 states east of the Mississippi, from Michigan and Wisconsin in the north to Tennessee and North Carolina in the south. In the summer of 2013, Kyle and Mike extended their collections west of the Mississippi River to the Rocky Mountains and along the Gulf Coast for a total of 29 states. The sample now includes 2363 fish, mostly bluegills. He will soon be starting on the PCR and sequencing. We will see what happens.

His collecting partner, Mike Zimmermann, gathered *Helisoma anceps* to obtain echinostome metacercariae from the same localities; his total count of *H. anceps* exceeded 9000. He will be looking at the population genetics and biology of the trematode. Their research was arranged so that they could help each other with their sampling. Mutual cooperation is a key element in my thinking with respect to graduate student training.

Trachemys s. scripta had been under investigation by Whit for just a couple of years before Joe Bourque began his work on the turtle parasites at SREL. As I mentioned earlier, Joe and I met when he came up to Gull Lake and KBS to take my parasitology course. When he finished his undergraduate degree at the University of

Illinois, he opted to come to Wake Forest, where he obtained an MS degree in 1969 and his PhD in 1974. His fieldwork research at SREL (Bourque, 1974) covered 1970–1972 and included 324 turtles from five sites in river swamps and nine different ponds, all of which were influenced by natural temperatures. Also included were a thermally altered creek and pond. Overall, 10 species of helminth parasites were found in the turtles. The prevalence of *Camallanus* sp., a nematode, was the highest (90%); *Spironoura* sp., another nematode, was a close second (73%). Collectively, the prevalence of *Neoechinorhynchus* spp. (Acanthocephala) was also at 73% (included in this total were four identifiable species of *Neoechinorhynchus* and one unknown species). There were three species of trematodes, the most prevalent being *Telorchis medius* at 27%.

Most of Joe's work focused on examining the relationship between thermal loading (=stress) and the population biology of the acanthocephalan helminths. While we do not know for certain, it is highly probable that all four species of *Neoechinorhynchus* have very similar life cycle patterns, with ostracods as the primary intermediate hosts. On reviewing the reports of Bourque (1974), Bourque and Esch (1974), Esch et al. (1980a, 1980b), and Jacobson (1987), a couple of distinct biological characteristics stood out. In the first place, diet and age of most turtle species are clearly coupled with the recruitment of certain helminth parasites. Many turtle species, including *T. s. scripta*, are carnivores as juveniles, feeding primarily on insects, snails, and so on. Adults of the same species then become herbivorous with the passage of time. For aquatic and semiaquatic species of turtles, algae of various kinds are their food of choice. However, in the process, they will also accidentally encounter algal epifauna species such as ostracods, which are the primary intermediate hosts for *Neoechinorhynchus* species in aquatic ecosystems. This pattern accounts for the gradual increase in abundance of acanthocephalans as the turtle's age. It also should be noted that the life spans of most acanthocephalans are probably no longer than a year. Since adults of *Neoechinorhynchus* spp. are present throughout the year in turtles, this suggests that their recruitment and turnover are constant over time. The only thing that might diminish the numbers of these parasites in the turtles would be a reduction in food consumption during the winter months when turtle metabolism is reduced as a result of declining water temperatures.

Parasite diversity in turtles appeared to be related to habitat stability on the SRP. For example, the greatest helminth diversity was associated with the river swamp habitats, whereas the smallest diversity was associated with a small pond receiving thermal effluent. There is clearly a wide range of habitat varieties on the SRP, and this is reflected in the wide range of parasite community diversities as well.

My old friend in England, Clive Kennedy, and his colleagues, Al Bush and John Aho (Kennedy et al., 1986), identified several factors that contributed to the establishment of interactive/isolationist communities in piscine and avian hosts. They included host vagility, endothermy, complexity of the gastrointestinal tract, and breadth of the host diet. Using the same factors, Esch et al. (1990) developed a summary regarding the complexity and patterns of helminth community structure in turtles using information gathered by Joe, Whit, John, Al, Clive, and other investigators over the years:

- Chelonians, including sliders, are quite vagile, and with increased vagility, there should also be increased exposure to potential intermediate hosts or to parasites with direct life cycles.
- Parasites, in much greater abundance, have been described from semiaquatic emydid turtles such as *T. s. scripta* as compared to terrestrial species, and even though older *T. s. scripta* are strict herbivores, they nonetheless possess a rich acanthocephalan fauna, which is related to the accidental consumption of ostracods that are part of the epifaunal community of their algal food source.
- Body temperatures in turtles are elevated over amphibians and fish especially while basking, and consequently, their food consumption should also be greater, thereby increasing their chances for greater exposure to parasite-infected intermediate hosts, and other transmission stages.
- Turtle enteric systems are more complex than those of both amphibians and fish, and therefore, the opportunity for more helminth colonization should be reflected in greater parasite community diversity in turtles, and this certainly appears to be the case in *T. s. scripta* on the SRP site.

Kym Jacobson came to Wake Forest in 1984 from the University of Nevada–Reno to work on her master's degree. After we had some discussion, she decided to follow up on some of Joe Bourque's research on the helminth fauna of slider turtles in Par Pond. One of her goals was to examine the linear and circumferential distribution of acanthocephalans along the intestine. Her findings revealed that the four *Neoechinorhynchus* species completely overlapped in their distribution, indicating that competitive interactions among these acanthocephalans were not important in structuring the infracommunities in *T. s. scripta*. However, Kym did not particularly like the killing and necropsy part of doing parasite ecology, so she switched into Ray Kuhn's lab and undertook some immunology research to obtain her PhD. Regrettably, her postdoc at the Mayo Clinic was a terrible experience. However, she subsequently managed to secure a postdoctoral position at a NOAA lab in Oregon, where she switched back into the ecological aspects of parasitology. She has since become a real leader in the marine facets of the field, and I am quite proud of her tenacity and ingenuity.

My students have done some interesting research down at the SRP, and our collaboration with Whit Gibbons has been highly productive over the years. Regrettably, our work down there has tapered off over time. I have one more story to tell about a line of study by another student, Terry Hazen, but this will come in another chapter. Also, unfortunately, the US Department of Energy has recently cut all funding for SREL. It is my understanding that only a skeleton crew is continuing—too bad, because they (we) had a really wonderful run while the lab was fully operational.

References

Bourque, J.E. 1974. Studies on the population biology of helminth parasites in the yellow-bellied turtle. Ph.D. dissertation. Wake Forest University, Winston-Salem, North Carolina, 105 p.

Bourque, J.E., and G.W. Esch. 1974. Population ecology and species diversity of parasites in *Psuedemys s. scripta* from thermally altered and natural aquatic communities. *In* Thermal ecology, G. Esch, and R. McFarlane (eds.). AEC Symposium Series, (CONF-730505). Springfield, VA, p. 551–561.

Esch, G.W., and J.W. Gibbons. 1967. Seasonal incidence of parasitism in the painted turtle, *Chrysemys picta marginata* Agassiz. *Journal of Parasitology* **53**: 818–822.

Esch, G.W., J.W. Gibbons, and J.E. Bourque. 1975. An analysis of the relationship between stress and parasitism. *American Midland Naturalist* **93**: 539–553.

Esch, G.W., J.W. Gibbons, and J.E. Bourque. 1980a. Species diversity of helminth parasites in *Chrysemys s. scripta* from a variety of habitats in South Carolina. *Journal of Parasitology* **65**: 624–638.

Esch, G.W., J.W. Gibbons, and J.E. Bourque. 1980b. The distribution and abundance of helminth parasites in *Chrysemys s. scripta* from a variety of habitats on the Savannah River Plant in South Carolina. *Journal of Parasitology* **65**: 633–638.

Esch, G.W., D.J. Marcogliese, T.M. Goater, and K.C. Jacobson. 1990. Aspects of the ecology and evolution of helminth parasites in turtles: A review. *In* Life history and ecology of the slider turtle, J.W. Gibbons (ed.). Smithsonian Institution Press, Washington, DC, p. 299–307.

Eure, H.E. 1974. Studies on the effects of thermal effluent on the population dynamics of heminth parasites of the largemouth bass, Micropterus salmoides. Ph.D. dissertation. Wake Forest University, Winston-Salem, North Carolina, 95 p.

Eure, H. 1976. Seasonal abundance of *Proteocephalus ambloplitis* (Cestoidea: Proteocephalidea) from largemouth bass living in a heated reservoir. *Parasitology* **73**: 205–212.

Fischer, H., and R.S. Freeman. 1969. Penetration of parenteral plerocercoids of *Proteocephalus ambloplitis* (Leidy) into the gut of smallmouth bass. *Journal of Parasitology* **55**: 766–774.

Hunter, G.W. 1928. Contributions to the life history of *Proteocephalus ambloplitis* (Leidy). *Journal of Parasitology* **14**: 229–243.

Jacobson, K.C. 1987. Infracommunity structure of enteric helminthes in the yellow-bellied slider, *Trachemys scripta scripta*. Master's thesis. Wake Forest University, Winston-Salem, North Carolina, 46 p.

Kennedy, C.R., A.O. Bush, and J.M. Aho. 1986. Patterns in helminth communities: Why are birds and fish different? *Parasitology* **93**: 205–215.

5 Development of Some Conceptual Notions

Reader, look, not at his picture, but his book.
On the Portrait of Shakespeare, Prefixed to the First Folio,
Ben Jonson (1573–1637)

Research on the ecological aspects of parasitology began as natural history investigations long ago and transitioned to natural science by the early 1960s (Wharton, 1962). Most of the work in parasitology during the late 19th and early 20th century was focused on standard species descriptions, morphology, and life cycles. Rudolph Leuckart and Patrick Manson were among the leaders in those years. Among other things, Leuckart established that larvae of the tapeworms *Taenia saginata* and *Taenia solium* were confined to cattle and swine, respectively, with humans as definitive hosts (although larval *T. solium* can also infect humans = cysticercosis). Manson is associated with elephantiasis and *Wuchereria bancrofti* and, in 1878, was the first to determine that mosquitoes were vectors for a helminth parasite. Overall, his contributions and impact were so significant that he is frequently referred to as the "father" of tropical medicine. A.P.W. Thomas and R. Leuckart, although working independently, are given joint credit for elucidating the life cycle of *Fasciola hepatica* in 1882–1883, a first for digenetic trematodes.

Perhaps the most significant success of that era was achieved by Ronald Ross, who, in 1897, successfully resolved most of the life cycle of a *Plasmodium* species from birds while working in India. He had been stimulated by Manson's hypothesis

Ecological Parasitology: Reflections on 50 Years of Research in Aquatic Ecosystems,
First Edition. Gerald W. Esch.
© 2016 John Wiley & Sons, Ltd. Published 2016 by John Wiley & Sons, Ltd.

that mosquitoes were the vectors for the dread parasite that caused malaria. Ross' success led to his receipt of a Nobel Prize in 1902. Unfortunately, Fritz Schaudinn mistakenly reported in 1903 that the *Plasmodium* sporozoite (a developmental stage in the life cycle) immediately entered red blood cells on being injected into the vertebrate host by a mosquito. His "supposed" observation was not corrected for nearly 50 years. Thus, it was not until 1948–1949 that Henry Shortt and P.C.C. Garnham cleared up that mess when they successfully determined that both *Plasmodium vivax* and *Plasmodium falciparum* first established in the liver before moving into the blood vascular system.

The early discoveries were followed into the 20th century by a great many successful field studies involving the life cycles of protozoans, cestodes, trematodes, and nematodes. People like Wendell Krull, Horace Stunkard, William Walter Cort, T.W.M. Cameron, and Harold Manter were some of the notable North American field parasitologists of that era.

In 1941, V.A. Dogiel, the great Russian parasitologist, published the first book dealing with ecological parasitology, titled *A Course in General Parasitology*. Subsequent to his death in 1955, two of his former students, Y.I. Polyanski and E.M. Kheisin, undertook an effort to revise the second edition of Dogiel's first tome. In doing so, they changed the title of the original book to *General Parasitology* while retaining Dogiel as the lone author. If one examines the content of the final Dogiel text, the ecological thrust is clear. In their foreword, Polyanski and Kheisin stressed that the book "… focused attention on many theoretical problems of parasitology and for the first time presented a new direction in parasitological investigations, broadly developed in the Soviet Union, and known as 'ecological parasitology.'"

Their assertion regarding the launch of ecological parasitology, vis–à–vis Dogiel, cannot be challenged, although I am certain that most current parasite ecologists would consider the book's thrust as natural history (considered as observational), not natural science (considered as empirical). My assessment of Dogiel's significant contribution is certainly not intended to be disparaging. However, I think most of us would agree that the move into modern ecological parasitology began with the three papers by John Holmes that he published in the *Journal of Parasitology* in 1961–1962. The quantitative focus of field parasitology came with Harry Crofton's modeling papers published in *Parasitology* in 1971–1972. Subsequently and without question, the most sophisticated mathematical modeling papers dealing with the epidemiology of infectious diseases were published by Anderson and May (1979) and May and Anderson (1979).

The work by Holmes (1961, 1962a, 1962b) was stimulated by Clark P. Read's (1951) paper in which he described what was termed the "crowding effect." In reality, Holmes' investigations were testing the so-called competitive exclusion principle devised by Grinnell in 1904, that is, that "… two species of approximately the same food habits are not likely to remain long evenly balanced in numbers in the same region. One will crowd out the other." In a fairly recent review, Bush and Lotz (2000) praised Holmes' work by stating that he (Holmes)

> provided the most conclusive experimental evidence for interspecific competition since Gause's (1934) elegant experiments on competition between *Paramecium caudatum* and *Paramecium aurelia* and Parks' (1948) equally impressive paper on competition between *Tribolium confusum* and *Tribolium castaneum*. To our knowledge, Holmes' papers were the first to demonstrate, experimentally, interspecific competition where the competitors were phylogenetically distant (different phyla in this case [Platyhelminthes and Acanthocephala]).

Holmes' (1961, 1962a, 1962b) efforts were to later stimulate a significant push toward understanding various aspects of parasite community ecology (Bush and Holmes, 1986a, 1986b; Esch et al., 1988; Sousa, 1990; Marcogliese and Cone, 1991; Sukhdeo, 1991; Lotz and Font, 1994; Sukhdeo and Sukhdeo, 1994; Janovy et al., 1995; Poulin, 1995; Rohde, 2005).

About 10 years after Holmes' papers were published, Harry Crofton (1971a, 1971b) introduced a sophisticated quantitative approach to parasitology. Crofton had become disenchanted with the manner in which parasitism was being defined. In his judgment, the methods were too qualitative, and he preferred a more quantitative slant. After some preliminary attempts, he became convinced that quantitative methodology could best be applied through use of the negative binomial model. He referred to it as a "fundamental model" that was paramount in describing a parasite frequency distribution among its hosts. Parasite populations that can be described by the model have a hypothetical basis for inferring the manner in which they are able to arise. As such, the hypotheses can then be tested and either supported or rejected. An important aspect of his approach included the notion that the frequency distributions of many parasite populations are aggregated, that is, they are overdispersed or contagious. It must be noted that not all parasite populations can be described by modeling of this sort, because they reproduce asexually in a host, creating large numbers of cloned offspring.

Elsewhere, I wrote about a conversation (Esch, 2004) between Clive Kennedy and Crofton when they were traveling home from London by train after a British Society for Parasitology Council meeting and prior to the publication of the latter's two seminal papers in which he described the concept of overdispersion/aggregation and then applied it using mathematical modeling. Crofton told Clive that one of the questions he was trying to answer was why the frequency distributions of so many helminth infrapopulations were decidedly overdispersed. He had been struck by his observations that most of the sheep in a flock he was studying had relatively small numbers of the parasitic nematode and that only a few individuals had very large numbers. Because of his critical insight, we now know that most helminth species in host populations are distributed in the same way.

The research of Crofton in the early 1970s opened the proverbial "door" for a great many investigators into the classical aspects of parasite population ecology and epidemiology (Anderson and May, 1979; May and Anderson 1979; Croll and Ghadarian, 1981; Dobson, 1985; Haswell-Elkins et al., 1988; Holland et al., 1988; Esch et al., 1997; Kightlinger et al., 1998).

As implied in the previous chapter, the PhD work of Joe Bourque and Herman Eure had encouraged interaction between Whit Gibbons and I in a number of ways. Even before Joe and Herman showed up at Wake Forest, however, Whit and I had been talking about several ideas regarding stress and host–parasite interactions but in a totally different way from the "general adaptation syndrome" developed by Hans Selye back in the 1950s.

For example, in 1969, Dr. Glenn Noble of the California Polytechnic University in San Luis Obispo, California, contacted me. He asked if I would be interested in participating in a colloquium on stress and parasitism that he was organizing for the Second International Congress of Parasitology in 1970 in Washington, DC. When he invited me, I am certain that he wanted my participation to focus on the classic approach of Hans Selye and the involvement of corticosteroids such as cortisone. When he asked, I said yes, but only if I could include my friend, and colleague, Whit Gibbons, as a collaborator, and Glenn agreed. Whit and I had been toying with an idea that was to take stress and parasitism out of the ordinary and examine it from the perspective of an individual host and its parasites, as well as to extend it to host populations and ecosystems. Because of elevated water temperatures in many localities on the Savannah River Plant (SRP) site, we believed we were dealing not only with individual hosts and their parasites but with host and parasite populations and communities within the context of thermally altered, or stressed, ecosystems.

So, Whit and I sat down and wrote an abstract focusing on some very preliminary work at SRP relating turtles, their parasites, and stress (Joe Bourque had just started as a PhD student). Unfortunately, Whit was unable to attend the DC meeting, so I had to hold our position alone. At the colloquium, I found myself with a bunch of Selye stress traditionalists, and I do not think that most of them had any idea what Whit and I were talking about. After the meeting in DC, Whit and I kept the conversation going for several more years. In 1973, Whit and I asked Robert MacIntosh, who was editor of the *American Midland Naturalist*, if he would consider publishing a review dealing with our ideas connecting stress and parasitism, and he agreed to consider the manuscript. So, during a visit to the Kellogg Biological Station (KBS) that summer, Whit and I began to put together some of our ideas on paper.

As noted earlier, perhaps the pioneering scientist in developing the concept of stress and stress theory was Hans Selye. His conceptualization included the idea that stress is "… the sum of all physiological responses by which an animal attempts to maintain or re-establish a normal metabolism in the face of a physical or chemical force" (Selye, 1950). Our main problem with this approach was that it was too narrow and that it was focused at the individual level. Our thinking was that stress could operate at all levels of biological organization and on both plants and animals, parasites included (Esch et al., 1975). Accordingly, in our approach, we attempted to concentrate on the idea that "... stress is the effect of any force which tends to extend any homeostatic or stabilizing process beyond its normal limit, at any level of biological organization."

The response to our paper was actually very good. Although stress and parasite biology were intended to be the real focal point of our effort, we were pleased to find that a really positive response was to another idea, or concept, that we introduced in the stress paper.

I can clearly recall an evening when Whit and I sat alone in the dining hall at KBS discussing the population concept as it applied to many kinds of parasites. Whit is not a parasitologist and I was trying to explain to him that the population biology of parasites couldn't be considered in the same manner as that for a free-living population. In fact, they are very different. A free-living population is normally defined as a group of organisms of the same species occupying a given space. But, in our paper (Esch et al., 1975), we posed the important and relevant question, "Do all members of a given parasite species within a single host constitute a population, or should all members of a species in all hosts within a given ecosystem be considered a population?" It was emphasized then, and here again, that this is not just a matter of semantics. It was our position that a population of free-living organisms increases through birth or immigration or both. In contrast, within a given host, the number of most helminth parasites can increase only through immigration (= recruitment). For most protozoans and some helminths, populations (e.g., in their first intermediate hosts) increase by both birth (asexual reproduction) and immigration. In effect, our stance was that we were dealing with an unusual kind of population dynamics for some parasites, although certainly not all. I cannot honestly say which of us came up with the new terminology, but it was at KBS that evening when we initially developed the ideas of infra- and suprapopulations. The former we defined as "... all individuals of a single parasite species within an individual host." The suprapopulation we defined as "... including all parasites of a given species, in all stages of development, within all hosts of an ecosystem." We then went on to discuss parasite population biology and stress within the context of infra- and suprapopulations.

Expansion of these ideas for our paper came quickly after we arrived at a resolution for the new population concept. Whit had finished PhD by then and was in a tenure-track position at SREL, while Joe Bourque was still working on his PhD at SREL. I remember the last night of writing the Esch–Gibbons–Bourque paper (1975). I was in my office at the research/teaching building at the KBS. I can even remember that toward midnight there was a severe thunder and lightning storm. Although it had taken us nearly 7 years of thinking, and talking, and writing, to reach our conclusions, our ideas seemed to come to a head suddenly. The concepts generated in that paper caught on in the ecological parasitology literature quickly and continue in use to the present. Whit and I were truly flattered when Al Bush and John Holmes saw fit to expand these ideas to community ecology as well, for example, infracommunities and supracommunities (now considered as compound communities), in Al's dissertation and subsequent papers dealing with parasites in lesser scaup (Bush and Holmes, 1986a).

Not long afterward, Austin MacInnis (then editor of the *Journal of Parasitology*) asked Dr. Elmer Noble (president of the American Society of Parasitologists (ASP) and Glenn Noble's brother) to appoint a committee (Leo Margolis, Gerald Esch, John Holmes, Armand Kuris, and Gerhard Schad) to consider ecological terminology and prepare a recommendation for what should and should not be used in our literature. A few years after our report was published (Margolis et al., 1982), we were notified by Thomson Reuters Publishing Co. (the old Institute for Scientific Information, or ISI) that the paper had become a "citation classic." In a review paper celebrating the centennial anniversary of the *Journal of Parasitology*, Scholz and Choudhury (2014) reported that the Margolis et al. (1982) paper was subsequently cited approximately 1500 times prior to 1997.

When I became editor of the *Journal of Parasitology*, I asked Al Bush to assemble another committee that included Kevin Lafferty, Jeff Lotz, and Al Shostak and provide an update. When Whit, Joe, and I described the infra- and suprapopulation concepts in 1975, we had overlooked several important issues, but Bush et al. (1997) corrected the problem. Thus, they referred "...to all of the individuals of a specified life history phase at a particular place and time" as a component population. As I mentioned earlier, Al Bush and John Holmes (1986a) extended our ideas regarding parasite populations to the community realm. Their idea was that an infracommunity parallels an infrapopulation as all parasite species within a single host. Similarly, a component population is analogous with a component community and that suprapopulations and compound (or supra-)communities are also parallel. Another note of interest is that according to Scholz and Choudhury (2014), the Bush et al. (1997) paper had been cited approximately 2300 times since its publication in the *Journal of Parasitology*.

In the fall of 1974, Terry Hazen (a grad student mentioned previously in connection with *Caulanthus cooperi* in an earlier chapter) and I were invited by Joe Schall to participate in an American Association for the Advancement of Science symposium in San Francisco. The topic was ecological parasitology. Another participant in the symposium was Al Bush. By that time, I had met Al at several meetings of the ASP, and we had by then become very good friends. Neither Al nor I was "good fliers," so since we were leaving on separate fights the same day, we decided to take a taxi together to the airport and do some imbibing (as a way of building courage) prior to leaving for home. We had to wait a few hours, and as was typical for us, our discussion turned to our research. During our ecological and parasite meandering, we came upon a question that we felt needed resolution. Our problem was focused on colonization strategies, especially among parasites that are associated with aquatic ecosystems.

Al and I continued to meet at various intervals over the ensuing years. Each time, we would discuss the same issue, and we seemed to be getting closer to a resolution.

In fact, we had even developed the terminology for two colonization ideas, autogenic and allogenic, before Clive and John Aho joined us the year before the ASP met at Wake Forest University in 1988. Clive was the old friend who I talked about in the first chapter and one of the people who set me on the ecology track back in 1971–1972.

John Aho, as previously described, was a former student of mine. He then went over to the University of Exeter to do his PhD with Clive, and I was even invited by John and Clive to be John's external examiner. Ann accompanied me and we recall the trip with great fondness. We stayed on the campus in a beautiful visitor's house, where each morning we were served an elegant English breakfast by the house staff. John's research site was in the River Swincombe up on Dartmoor in Devon. We enjoyed our trip up on the barren moor and were anticipating our ride down to Slapton Ley to visit one of Clive's favorite research sites.

In a historical context, I was particularly fascinated by Slapton Sands (the beaches just to the south of Slapton Ley) and the cliffs to the east of the lake alongside the English Channel. I had learned that the beaches were used to rehearse for the Allied landing at Normandy in WWII. The cliffs, close by the Slapton Sands, were employed by American Rangers to practice their landing at Point du Huc during the same invasion. The evening before driving down to Clive's research site, Ann and I were watching the BBC on television. We saw in the news that an American Sherman tank had been dedicated (November 15, 1987) at Slapton Ley to the memory of about 800 American soldiers and sailors who had lost their lives on the night of April 28, 1944. It seems that several German E-boats (similar to American PT boats) had penetrated into a group of American LST's (landing ship, tank) carrying American soldiers who were being transported, along with their equipment, to rehearse their landing at Normandy. Most of the local residents in the vicinity had been relocated during the months preceding the Normandy invasion and were unaware for a long period of time of the WWII action at Slapton Sands. They were partly responsible for promoting the dedication ceremonies. Our trip to Slapton Ley was well worth the time. To anyone with an interest in history or parasite ecology, it is a very impressive place.

Just before the 1988 ASP meeting at Wake Forest, Al, John, Clive, and I sat and talked about the idea that Al and I initially germinated and then invited Clive and John to work with us. Actually, Clive did more than just talk about it. As our conversation was drawing to a close, Clive volunteered that he had some "unused" data from Great Britain that coincided with our ideas and that it would be easier for him to write the summary publication than any of the other three of us. So, even though I am the senior author, Clive was the primary contributor for the paper.

In our focus on parasite colonization strategies in aquatic ecosystems, two categories of helminths are recognized, that is, autogenic species, which mature sexually in fish or other vertebrates constrained entirely to aquatic systems, and allogenic species, which mature in vertebrates that are transient visitors to aquatic systems and, consequently, possess greater colonization potential. Our strategic premise was based on the apparent stochastic nature of freshwater fish helminth communities, which illustrates what appears to be the seemingly inconsistent and unpredictable occurrence and allocation of fish parasites. Clive's data were based on three groups of fishes, that is, anadromous, catadromous, and stenohaline. The first group includes fish that migrate upstream from marine systems into freshwater in order to spawn. Catadromous species do just the opposite; they migrate from freshwater to marine habitats to reproduce. Stenohaline species, on the other hand, are limited in their tolerance to wide ranges of salinity.

Kennedy's data revealed that salmonids tend to be dominated by autogenic parasite species. He noted that these autogenic species accounted for a great deal of the similarity within, and between, various study locations. Cyprinids are just the opposite. They are dominated by allogenic species, which also account for much of the similarity within, and between, sites. Anguillids are in between, with neither autogenic nor allogenic species being dominant. Their findings indicated that

> Our conclusions relating to the different contributions of autogenic and allogenic species to community structure in the three groups are novel and unexpected, and are not simply logical consequences of the dispersal abilities of the helminths themselves. An understanding of colonization strategies, therefore, including the separation of autogenic and allogenic species and recognition of the different roles of both transient/resident and euryhaline/stenohaline host species, provides important clues for evaluating the stochastic nature of parasite community structure. Esch et al. (1988)

Unquestionably, Dogiel, Holmes, Crofton, Kennedy, Anderson, May, Bush, and others played significant roles in developing the conceptual core for modern parasite ecology in the 1960s and 1970s. Many of the ideas and thinking that followed were grounded on their distillations of how parasites and their hosts networked within the framework of population and community ecology as ultimately derived from the study of free-living species. The advancements and changes that have occurred in ecological parasitology in the subsequent years are premium reflections of the innovative work accomplished by these early investigators.

References

Anderson, R.M., and R.M. May. 1979. Population biology of infectious diseases: Part I. *Nature* **280**: 361–367.

Bush, A.O., and J.C. Holmes. 1986a. Intestinal helminths of lesser scaup ducks: Patterns of association. *Canadian Journal of Zoology* **64**: 132–141.

Bush, A.O., and J.C. Holmes. 1986b. Intestinal helminthes of lesser scaup ducks: An interactive community. *Canadian Journal of Zoology* **64**: 142–152.

Bush, A.O. and J.M. Lotz. 2000. The ecology of crowding. *Journal of Parasitology* **86**: 212–213.

Bush, A.O., K.D. Lafferty, J.M. Lotz, and A.W. Shostak. 1997. Parasitology meets ecology on its own terms: Margolis et al. revisited. *Journal of Parasitology* **83**: 565–583.

Crofton, H.D. 1971a. A quantitative approach to parasitism. *Parasitology* **62**: 179–193.

Crofton, H.D. 1971b. A model for host-parasite relationships. *Parasitology* **62**: 343–364.

Croll, N.A., and E. Ghadarian. 1981. Wormy persons: Contributions to the nature and patterns of overdispersion with *Ascaris lumbricoides, Ancylostoma duodenale, Necator americanus,* and *Trichuris trichiura. Tropical and Geographical Medicine* **33**: 241–248.

Dobson, A.P. 1985. The population dynamics of competition between parasites. *Parasitology* **91**: 317–347.

Esch, G.W. 2004. Parasites, People, and Places. Cambridge University Press, Cambridge, 235 p.

Esch, G.W., J.W. Gibbons, and J.E. Bourque. 1975. An analysis of the relationship between stress and parasitism. *American Midland Naturalist* **93**: 339–353.

Esch, G.W., C.R. Kennedy, A.O. Bush, and J.M. Aho. 1988. Patterns in helminth communities in freshwater fish in Great Britain: Alternative strategies for colonization. *Parasitology* **96**: 519–532.

Esch, G.W., E.J. Wetzel, D.A. Zelmer, and A.M. Schotthoefer. 1997. Long-term changes in parasite population and community structure: A case history. *American Midland Naturalist* **135**: 369–387.

Haswell-Elkins, M.R., D.B. Elkins, K. Manjula, E. Michael, and R.M. Anderson. 1988. An investigation of hookworm infection and reinfection following mass anthelminthic treatment in the South Indian fishing village of Vairavankuppam. *Parasitology* **96**: 565–577.

Holland, C.V., D.L. Varen, D.W.T. Crompton, M.C. Neshiem, D. Sanjur, I. Barbeau, K. Tucker, J. Tiffany, and G. Rivera. 1988. Intestinal helminthiases in relation to socioeconomic environment in Panamanian children. *Social Science and Medicine* **26**: 209–213.

Holmes, J.C. 1961. Effect of concurrent infections on *Hymenolepis diminuta* (Cestoda) and *Moniliformis dubius* (Acanthocephala). I. General effects and comparison with crowding. *Journal of Parasitology* **47**: 209–216.

Holmes, J.C. 1962a. Effect of concurrent infections on *Hymenolepis diminuta* (Cestoda) and *Moniliformis dubius* (Acanthocephala). II. Effects on growth. *Journal of Parasitology* **48**: 87–96.

Holmes, J.C. 1962b Effect of concurrent infections on *Hymenolepis diminuta* (Cestoda) and *Moniliformis dubius* (Acanthocephala). III. Effects in hamsters. *Journal of Parasitology* **48**: 97–100.

Janovy, Jr., J., R.E. Clopton, D.A. Clopton, S.D. Snyder, A. Efting, and L. Krebs. 1995. Species density distributions as null models for ecologically significant interactions of parasite species in an assemblage. *Ecological Modeling* **77**: 189–196.

Kightlinger, L.B., J.R. Seed, and M.B. Kightlinger. 1998. *Ascaris lumbricoides* intensity in relation to environmental, socioeconomic and behavioral determinants of exposure to infection in children from Southeast Madagascar. *Journal of Parasitology* **84**: 480–484.

Lotz, J.M., and W.F. Font. 1994. Excess positive associations in communities of intestinal helminths of bats: A refined null hypothesis and a test of the facilitation hypothesis. *Parasitology* **103**: 127–138.

Marcogliese, D.J., and D.L. Cone. 1991. Importance of lake characteristics in structuring parasite communities of salmonids from insular Newfoundland. *Canadian Journal of Zoology* **69**: 2962–2967.

Margolis, L., G.W. Esch, J.C. Holmes, A.M. Kuris, and G.A. Schad. 1982. The use of ecological terms in parasitology (report of an ad hoc committee of the American Society of Parasitologists). *Journal of Parasitology* **68**: 131–133.

May, R.M., and R.M. Anderson. 1979. Population biology of infectious diseases. II. *Nature* **280**: 455–461.

Poulin, R. 1995. Hairworms (Nematomorpha: Gordioidea) infecting New Zealand short-horned grasshoppers (Orthoptera: Acridiidae). *Journal of Parasitology* **81**: 121–122.

Read, C.P. 1951. The crowding effect in tapeworm infections. *Journal of Parasitology* **37**: 174–178.

Rohde, K. 2005. Marine parasitology. CABI Publishing, Wallingford, 497 p.

Scholz, T., and A. Choudhury. 2014. Parasites in freshwater fishes in North America: Why so neglected? *Journal of Parasitology* **100**: 26–45.

Selye, H. 1950. Stress and the general adaptation syndrome. *British Medical Journal* **1**: 1383–1392.

Sousa, W.P. 1990. Spatial scale and the processes structuring a guild of larval trematode parasites. *In* Parasite communities: Patterns and processes, G.W. Esch, A.O. Bush, and J.M. Aho (eds.). Chapman and Hall, London, p. 41–68.

Sukhdeo, M.V.K. 1991. The relationship between intestinal location and fecundity in adult *Trichinella spiralis*. *International Journal for Parasitology* **21**: 855–858.

Sukhdeo, M.V.K., and S.C. Sukhdeo. 1994. Optimal habitat selection by helminths within the host environment. *Parasitology* **109**: S41–S55.

Wharton, G.H. 1962. Future of systematics in parasitology. *Journal of Parasitology* **48**: 651.

6 The Pond: Part I

One is never entirely without the instinct of looking around.
One of the Human Kinks, Walt Whitman (1819–1892)

A new line of very productive research was begun quite by accident in the spring of 1982. The story begins with an undergraduate student by the name of Amy Crews (see Chapter 3) who took my course in parasitology at Wake Forest University. Toward the end of the semester, she came to my office and informed me that she wanted to go to graduate school and become a parasitologist. Amy was a very bright young woman, and I thought she could be successful. However, I honestly felt she would be more likely to succeed in the long run if she first had a real understanding about actually doing research. So, in order to gain some experience, I suggested that she work in the field during her senior year with another of my graduate students, Mike Riggs.

After some discussion the next year, I persuaded Amy that she should stay at Wake Forest and complete a master's degree before pursuing her PhD elsewhere. As it turned out, it is one of the best pieces of advice I have ever offered, not only for Amy but also for 10 more graduate students and me who followed her into the same small pond. The work conducted by these 11 students continued off and on for 31 years and has yielded 30 publications, with several of them dealing primarily with the biology of two species of trematode. All of this work was focused in a small, two-hectare pond (Figure 6.1) located adjacent to Belews Lake (North Carolina), a very large cooling reservoir operated by Duke Energy (more about this remarkable lake in Chapter 8).

After being accepted into our graduate program, Amy came to me late in her senior semester and asked if she could get an early start on her research—I agreed. I told her about the small impoundment located next to Belews Lake and that we

Ecological Parasitology: Reflections on 50 Years of Research in Aquatic Ecosystems,
First Edition. Gerald W. Esch.

Figure 6.1 Charlie's Pond in North Carolina. The pond was probably created during construction of the cooling reservoir (Belews Lake) in 1970. Our research first began in the pond during 1983 when Amy Crews and I began collecting snails.

should go there the next day and collect some snails, which we did. The snails were returned to the lab, where they were isolated in small plastic jars, half filled with aged tap water. It was her job to check the jars for cercariae the next day. Anyone who has ever collected snails and looked for cercariae knows that, most of the time, some trematode larvae will be released within 24–36 hours. Furthermore, in certain types of habitats, one can frequently predict the identity of some of the trematode species that will be present even before checking the jars for shedding cercariae.

The next day, this proved to be the case—at least for the most part. After carefully checking the snails, Amy came to my office and invited me to come to the lab. As we thought would happen, free-swimming cercariae were present, but a majority of the jars had something else entirely. At the bottom of most of the jars were scores, if not several hundreds, of tiny, round-shaped bodies, each with a small arm, or handle, protruding from one side, but we had no idea what they were (Figure 6.2). So, we pulled some of them into a Pasteur pipette, transferred them to a slide, and added a coverslip. I recall taking a probe and began to gently press the coverslip when, suddenly, there appeared to be a small, but impressive, explosion! Abruptly, a relatively long and apparently hollow tube had emerged from inside the round body, and simultaneously, something appeared to be "shot" through the tube to the outside (Figure 6.3). The analogy of a shell being fired through a cannon barrel is very

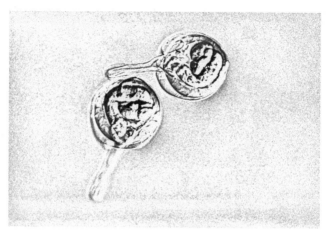

Figure 6.2   Two cercariae of *Halipegus occidualis*. The delivery tube and body of the trematode are still inside the tail. Photo courtesy of Derek Zelmer.

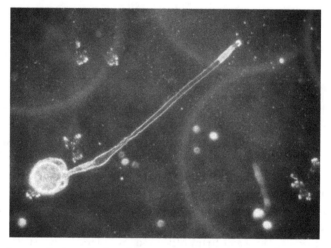

Figure 6.3   In an explosive episode, a delivery tube emerges, and simultaneously, the body of the cercaria is shot through the tube and into the hemocoel of an ostracod. Photo courtesy of Derek Zelmer.

accurate. Several times, we repeated the isolation of the round bodies and pressure on the coverslip, each time with the same result—it was a real show!

I returned to my office and Amy went searching for the identity of the round objects. It was not long before she returned with an answer. She discovered that we were observing the cercariae of *Halipegus* (=*Hal.*) *occidualis*, a hemiurid trematode (Figure 6.4), the adults of which occur under the tongues of green frogs, *Rana*

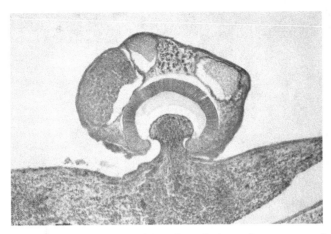

Figure 6.4   Using its ventral sucker, the adult *Hal. occidualis* attaches to tissue at the bottom of the mouth and under the tongue of green fogs, *Rana clamitans*. Photo courtesy of Eric Wetzel.

clamitans (Figure 6.5). We did not know it at the time, but Amy had found her master's thesis, as well as the master's theses or the PhD dissertations for ten more students.

On inquiring with Duke Energy officials, we learned that the small pond adjacent to Belews Lake had no name, probably because it was not a pre-Belews Lake farm pond. It was quite likely created when they constructed the power plant. The reservoir and pond were separated by a service road on top of a "riprap" dam. The pond has several interesting characteristics. First, it empties into Belews Lake, not the other way around. The pond is spring fed and excess water runs down and through a tunnel under the dam/service road and into the reservoir. Second, because it is spring fed, it was not significantly (or so we thought at the time) subjected to the drought conditions with which our area had contended regionally for the previous couple of years. Indeed, a great many farm ponds the size of our impoundment had completely dried up in our area of North Carolina! Third, the pond possessed several species of fish, dominated by mosquitofish and a few species of centrarchids, including black crappie, bluegills, and green sunfish. There are no longer any largemouth bass in the pond, the last having died, almost literally, in the hands of another graduate student, Tim Goater, who was to arrive on the scene in 1983. Tim was at the pond doing some work when he heard some splashing in a nearby cove. He walked over and pulled a flailing bass out of the water with his bare hands. We have not seen another bass in the pond since that day.

Once Amy began her work, we felt that we had to come up with a name for the pond. During our time at Savannah River Ecology Laboratory (SREL), I recalled that Whit Gibbons had given a name to nearly every cove in Par Pond, many of them referring to friends and colleagues at SREL, as well as to a couple of his own children. I liked the last idea, so we named it Charlie's Pond, after my youngest son, Charlie.

Figure 6.5   A green frog, *R. clamitans*. Photo courtesy of Tim Goater.

Typically, the percent of snails infected by both allogenic (AL = a fluke that completes its life cycle in a transient bird, mammal, or reptile) and autogenic (AU = a fluke that completes its life cycle within the confines of the body of water) trematodes in small farm ponds was relatively low, that is, less than 4%. In Charlie's Pond, this is what Amy observed for seven of the eight trematode species she initially encountered. There was an exception to this pattern, however. While collecting snails in the pond over a period of 1+ years, the prevalence of the hemiurid, *Hal. occidualis*, reached 100% a couple of times, and the overall annual prevalence the first summer was close to 60%, exceptionally high.

As an aside, hemiurid flukes most commonly occur in the stomachs/intestines of marine fishes. However, *Halipegus* spp. are parasites of frogs, usually found in the buccal cavity, esophagus, or Eustachian tubes. The life cycle of *Hal. occidualis* (Figure 6.6) was resolved by Krull (1935) (the experimental steps taken by Krull are described by Esch (2007)). The parasite's eggs are eaten accidentally by grazing *Helisoma* (=*Hel.*) *anceps*. After hatching and migrating to the hepatopancreas, the larvae develop into sporocysts, which, in turn, asexually produce rediae. Reproduction is prolific and rediae will spill over into the snail's gonads where they consume enough tissue to cause castration. Inside the rediae, cystophorous (nonswimming) cercariae are asexually produced, which are then shed from the snail. These cercariae are consumed by ostracods in which the explosive action described earlier force the bodies of the cercariae to enter the hemocoel of the

Figure 6.6 The life cycle of *Hal. occidualis*. Goater et al. (2014). Reproduced with permission of Cambridge University Press. Original drawing by Lisa Esch McCall.

microcrustacean. Dragonfly nymphs consume ostracods and the metacercariae take up residence in the lumen of the dragonfly nymphs. The cycle is completed when the ambush predator, *R. clamitans*, eats eclosing dragonfly nymphs. It should be noted here that cercariae in which the body of the fluke resides are sometimes referred to as cercariocysts (Macy et al., 1960).

Because of the exceedingly high prevalence of *Hal. occidualis* in Charlie's Pond, we agreed that Amy's work should focus on three aspects of the parasite's biology. First, she wanted to examine the seasonal changes in the population biology of *Hal. occidualis* in *Hel. anceps*. Second, she attempted to evaluate the fecundity characteristics of infected and uninfected snails to identify the extent of castration in the pond and, third, to histologically assess the effects of infection by *Hal. occidualis*. She consciously made the decision to focus on the snail–trematode interactions and not to deal with the frog–trematode combination, leaving that instead to Tim Goater who would start work in the pond later.

All of the data generated confirmed a strong seasonal pattern of infection by *Hal. occidualis* in the *Hel. anceps* population. There were peaks of infection in late spring and again in the fall. Changes in the size of snail shells indicated a 1-year life span in the pond, with snail senescence and population replacement occurring mostly in July, which corresponds nicely with the heaviest recruitment of the parasite in late spring and late summer into the fall. Snail fecundity was clearly affected by infection with the parasite and definitely related to mechanical damage (tissue consumption) by rediae.

In 1985, another player, Tim Goater, became involved with Amy and Charlie's Pond. Tim had arrived at Wake Forest from Al Bush's "shop" at Brandon University in Manitoba, Canada, in 1983 and completed his master's degree here in 1985 after working on salamander helminths in the Smokey Mountains of North Carolina. Tim then switched over to the *Hal. occidualis* system in Charlie's Pond to pursue his PhD and substantively expanded the start-up work of Amy Crews. He had a good thing going for him with his salamander study in the mountains. However, he chose to work with *Hal. occidualis* in green frogs because he knew that he would have to kill the mountain salamanders in order to enumerate infrapopulation sizes, something that he just did not have the "heart" to do.

While Tim was interested in snail–trematode interactions, he was much more focused on the frog's role in the population biology of the parasite. He had a perfect system in which to conduct this part of his research. Most of the traditional population studies performed using helminth parasites required a "kill and count" method, that is, kill the host and count the parasites. The problem with this approach is that the result produces just a snapshot of the parasite's population dynamics, leaving much to be desired. Because the host is dead, obviously nothing can be said about the rate at which parasites were recruited seasonally or what might have happened to the parasite infrapopulations after that point in time if the hosts had lived longer. The frog–*Hal. occidualis* system is totally different because these parasites can be enumerated without killing the frog host.

Tim collected frogs at night. After darkness set in, he would place his "windsurfer" board into the pond and load it with the supplies and gear necessary to collect. With a "miner's" lamp on his head, he would then wade the shoreline, listening for the call of "his" green frogs. On capture, the mouth of the frog would be opened, the tongue lifted, and the parasites counted (it was also easy to differentiate between immature and mature parasites). The frog would then be appropriately toe-clipped for identification and returned to the exact point of capture. The beauty of this system was that he could recapture frogs, which would allow him to follow the course of individual infrapopulations longitudinally. It also gave him a way to estimate the size of the frog population and any change that might occur during the course of study.

Serendipity in research has always fascinated me (Esch, 2007), and it ended up playing a significant role for Tim's research in Charlie's Pond. He wanted to know something about the population genetics of *Hal. occidualis* in Charlie's Pond. So, after he counted the parasites under the frogs' tongues, he would remove one parasite from each infrapopulation and place it into a vessel for freezing using dry ice, which was carried aboard his "windsurfer." All of the *Hal. occidualis* collected in this manner were shipped down to the SREL in Aiken, South Carolina. Upon arrival, a collaborator and population geneticist, Peg Mulvey, would then perform starch gel analyses on the samples (in today's world, the worms would be subjected to PCR and sequencing). We were also looking for a genetic difference between infected and uninfected snails and we did find a disparity (Mulvey et al., 1987). The data revealed a significant difference in genetic heterozygosity in two loci among infected and uninfected snails. It was suggested that "... factors conferring incompatibility may have been selected for in the *Hel. anceps* population within the pond," but no further work was directed at this facet of Tim's research—he had become focused on a couple of other findings associated with the snail–trematode system.

In one of Tim's collections, he found four parasites in the Eustachian tubes and made a note of the observation in his field book. A few weeks after he sent one of his batches down to the SREL, he received a telephone call from Peg. She informed him that, based on her starch gel electrophoresis analysis, four of the worms he had sent were different from the others included in the sample. On searching the literature, he discovered that green frogs also hosted *Halipegus* (=*Hal.*) *eccentricus* in the Eustachian tubes! The literature indicated that *Hal. eccentricus* used *Physa gyrina* as its first intermediate host. We knew *P. gyrina* had colonized the pond at some point after Amy began working there in 1983, but had made no effort to check them for parasites, at least to that point in time. The third intermediate hosts for *Hal. eccentricus* were said to be damselflies, which could have easily carried the parasite into our pond from another site. Tim checked the *P. gyrina* snails, and indeed, they were shedding *Hal. eccentricus*, which had also clearly colonized the pond without our knowledge.

A second important aspect of Tim's research had focused on the population biology of *Hal. occidualis* in the green frog (space limits the description of several other features of Tim's work; but see Goater (1989), Goater et al. (1990), and Fernandez et al. (1991)). As noted in an earlier chapter, overdispersion, or contagion, is characteristic for virtually all parasitic helminths during the adult stages of their life cycles, that is, variances among parasite infrapopulations are significantly greater than their mean intensities of infection. For example, if one examined humans for the roundworm, *Ascaris lumbricoides*, in southeastern Asia, approximately 20% of the villagers would carry about 80% of the parasites. This is referred to as an overdispersed, or contagious, frequency distribution. The bottom line for the idea is that a relatively small percentage of hosts carry a significantly larger percentage of the same parasite species within a given habitat.

Overdispersed frequency distributions can apparently be generated in a number of different ways. In the case of *Hal. occidualis* in green frogs, Tim hypothesized that contagion resulted from the frog's consumption of heavily infected dragonflies. For example, green frog #24 had six gravid parasites on first capture. Three weeks later, the same frog had 23 worms, 7 of which were immature. In other words, there were typically rapid increases in infrapopulation sizes throughout his work in the pond. Recruitment of the parasite by ostracods is a different matter. These tiny microcrustaceans ingest the cystophorous cercariae one at a time. Krull (1935), however, noted that the ostracods would feed voraciously on the motionless cystophorous cercariae and had counted as many as 16 parasites in a single host. It is no wonder that Dronen (1977) referred to some parasite species as being "biomagnified" as they move through the food web. Goater (1989) speculated that this sort of biomagnification could be viewed in the context of the evolution of complex parasite life cycles, for example, by "… adding additional hosts to a life cycle, many parasites have effectively increased their spatial and temporal dispersal and enhanced transmission success."

As just described, long-term studies of *Hal. occidualis* population dynamics in green frogs can be easily conducted simply by capturing the frogs, counting the parasites, toe-clipping the frogs, and then recapturing them. Some snails are also amenable to long-term studies since their infection status can be, in part, assessed easily by determining if they are shedding cercariae or not. Moreover, like green frogs, snails of some species can be marked, released, and recaptured so that growth characteristics of the snails and changes in parasite prevalence can be easily followed over time. One additional note should be added here. Not all snail species are suitable for these mark–recapture protocols because it is apparent that some snails move more than others. We learned this through direct experience. *Helisoma anceps* exhibits strong site fidelity, that is, their foraging ranges are quite small. Thinking we could apply the same research protocol to *P. gyrina*, several hundreds of the latter species were captured by Scott Snyder and then marked and released in a subsequent field experiment. On returning to the site of release, not a single snail was recaptured—almost all had moved away from the original release point! Scott's observation provided a great example of the importance of not only understanding the biology of the parasites but of their hosts as well.

During the fall of 1987, Al Shostak, Julie Williams, and Tim Goater decided to determine how well *Hel. anceps* infected with *Hal. occidualis* (and the parasite as well) would survive over winter in the pond. Altogether, 556 *Hel. anceps* snails were captured in a single cove of the pond (Fernandez and Esch, 1991a). The snails were brought to the laboratory where they were checked for infection status, that is, shedding cercariae or not. Each snail was painted with quick drying enamel paint and then inscribed with individual identification numbers using indelible ink and returned to the pond. Beginning in March of 1988 and continuing each month through June, these marked snails were recaptured and checked for infection status. Some very surprising results were obtained. For example, the data suggested that roughly 50% of the snails died over winter. The same dataset also indicated that infection by the parasites did not increase host mortality as we initially predicted. Histopathology and egg production experiments in the field (Crews and Esch, 1986, 1987) strongly suggested that gonad destruction by rediae of *Hal. occidualis* was complete, but it was not necessarily permanent. Thus, the Goater (1989) findings revealed that castration reversal had occurred in approximately 25% of the snails recovered in the spring. When Tim, Julie, and Al first began finding reproductively rejuvenated snails, they initially tried to blame each other, convinced that rejuvenation had not been a predictable phenomenon and that, instead, one of them had made a mistake in identifying infected and uninfected snails the previous fall. But mistakes were not made—it was determined that some of the snails simply lost their infections and became reproductively viable again!

Another observation worth noting was the absence of multiple infections in *Hel. anceps*, confirming results from the study by Crews and Esch (1986). This trend continued with work by Jackie Fernandez and Nick Negovetich. Based on their findings, it is clear that competitive interactions within the trematode community in Charlie's Pond occur and that a primary "battleground" is most likely inside *Hel. anceps*. The dominant trematodes in the snail infra- and component communities are species of echinostomes and *Hal. occidualis*, and both have rediae in their life cycles. It appears most likely that the presence of rediae in the parasites' life cycles and the rediae size are the crucial elements in success/failure of the competitive interaction between trematodes of different species inside snails. Other factors that may also contribute to diversity of the compound community include the temporal mode of parasite recruitment into the pond, physical characteristics of the pond itself, the abundance of *Chaetogaster* sp. (a predatory symbiont that occupies the mantle cavity of snails in the pond), and so on.

The next research step in the pond occurred with the arrival of Jackie Fernandez in 1987. Jackie is a native of Chile; she came to Wake Forest with a master's degree in hand from the University of Concepcion and at least 20 publications already to her credit. She had originally applied to do her graduate work with the late John Crites at Ohio State University. However, John was nearing retirement and recommended that she come and work in my lab. I remember driving to the airport with Dave Marcogliese and Tim Goater to pick her up. None of us knew quite what to expect, but I can say without any trepidation or hesitation that her contributions to our research were significantly greater than any of us could have wished for. Since we were in the middle part of our work with *Hal. occidualis* in Charlie's Pond, this is where she and I together decided she should go—the fit was perfect!

Jackie's research in the pond represents a comprehensive examination of the way in which infra- and component trematode communities operate in molluscan hosts within a confining pond habitat. It must be restated, however, that the system in which she conducted her research was completely dominated by one species, *Hal. occidualis*, with prevalences reaching as high as 100% more than once during a 15-month period and with recruitment of the parasite by snails at more than one collecting site in the pond. This qualification is of considerable importance since double patent infections in the pond were very rare (seven times from among 1485 infected *Hel. anceps* snails).

"Despite the dominance by a single species, spatial and temporal heterogeneity in the distribution and abundance of trematode infective stages indicate that not all snails have the same probability of becoming infected" (Fernandez and Esch, 1991b). She concluded that several factors were involved in affecting the heterogeneity in Charlie's Pond, that is, habitat structure, behavior of definitive hosts, the nature and timing of parasite recruitment, and the population dynamics of the snail hosts, including their seasonal mortality, reproduction, and life span. However, she also emphasized the importance of patch dynamics (Sousa, 1990) in creating the structure of the component community. She noted that a crucial factor in the patch dynamics concept is disturbance, as an annual reset mechanism. Begon et al. (1990) defined disturbance as "… any relatively discrete event in time that removes organisms and opens spaces which can be colonized by individuals of the same or different species." In the case of Charlie's Pond, the major reset mechanism is cohort turnover that takes place from late June to mid-July. In other words, when the snails die each summer, they take their parasites with them. However, the young snails recruited into the population each spring also begins recruiting a new component community, and in most years, it is very similar to the one it will replace by the time of snail death in midsummer.

Graduate students completed most of our work in Charlie's Pond. However, there was one postdoc, Dr. Al Shostak, who contributed significantly to the studies involving *Hal. occidualis*. Al came down to Wake Forest after he obtained his PhD with Dr. Terry Dick at the University of Manitoba where he completed a truly significant ecological study on the population biology of *Triaenophorus crassus* in northern pike. He was supported by fellowship from the Natural Sciences and Engineering Research Council of Canada. To me, Al was unlike a typical "postdoc." He was much more like a faculty colleague. Moreover, he is not only a first-class parasitologist/ecologist; he is also a really excellent statistician. The students that were working in the lab during his stay had the utmost respect for him as a person and as a scientist—so did I! He did some very interesting collaborative work with some of the students and also invited me to contribute in some of his independent research on *Hal. occidualis*. I was honored.

In addition to Al's overwintering collaboration work on *Hel. anceps* and *Hal. occidualis* with Tim and Julie, he was a leader on two studies that were primarily aimed at the survival and excystment behavior of the cystophorous cercariae produced by rediae of *Hal. occidualis*. The cercariae of most trematode species employ a "spatial" strategy for locating the next host in their life cycles. In other words, they swim. Even though the number of species subjected to experimentation is limited, there is every reason to believe that swimming behavior is not a random phenomenon. In other words, their swimming patterns are unquestionably influenced by several different kinds of environmental stimuli, chemical and physical, for example, light, gravity, current, mucous secreted by snails as they move across various substrata, and so on. The response to these stimuli places the cercariae in the vicinity of the next host in their life cycle. At this point, another set of cues specific for the snail species, sometimes of a chemical nature and others of a physical character, will aid the cercariae in identifying the precise location of the host. An excellent example of these phenomena is *Crepidostomum cooperi*, the allocreadiid fluke previously discussed in conjunction with eutrophication in Gull Lake.

However, swimming by cercariae is also a drain on the energy reserves of the parasite, especially since glycogen stores are finite and cercariae do not feed. The time limit for success is usually placed somewhere between 24 and 36 hours. If unsuccessful by the end of this time frame, the larval parasite dies. The prodigious reproductive capacity exhibited by most trematode species represents a tactic to compensate for the low success rate during this step in the parasite's life cycle.

However, cystophorous cercariae, like those of *Hal. occidualis*, do not follow a spatial strategy because they do not swim. Moreover, the body of *Hal. occidualis* inside the tail is anhydrobiotic (Goater et al., 1990), and accordingly, their metabolism at this stage is virtually nil. When they reach the ostracod's hemocoel, they begin to absorb water and are "turned on" metabolically. Because cystophorous cercariae of *Hal. occidualis* are metabolically inactive, they remain infective significantly longer (18–20 days) than their "swimming" counterparts. Accordingly, cystophorous cercariae employ a temporal strategy in host location—actually, their hosts find the cercariae, not the other way around.

Based on the early work by Crews and Esch (1986), it was determined that cercaria shedding by infected snails in Charlie's Pond is a seasonal phenomenon. When water temperatures are above 12°C, cercariae shedding will occur and cease when water temperature falls below 12°C. Given this phenomenon, two questions were raised by Al. First, what effect does water temperature have on transmission success once a cystophorous cercaria is shed from a snail? Second, is there a photocycle dependence on the release of cystophorous cercaria from infected snails? In preliminary studies, Macy et al. (1960) suggested that there was photocycle dependence, but Al wanted to confirm their observations.

Using data generated via laboratory experimentation and then by extrapolation curves developed by simulation predictions, he estimated that at 4°C, 90% of cercariae survived for 14 weeks and 50% survived for 30 weeks. In contrast, at 30°C, 90% survived for one and a half weeks and 50% survived for two and a half weeks. It was concluded that "… in temperate regions, with large seasonal variations in water temperature, the timing of cercarial release on a seasonal basis will have a large influence on their survival, and may therefore be an important component of population dynamics" (Shostak and Esch, 1990a).

Al's photocycle experiments clearly verified the Macy et al. (1960) report. There is a light/dark cycle, with more cercariae being shed at night than during the day. However, nocturnal emergence then raises questions regarding adaptive significance of the behavior. Three hypotheses to explain advantages in nocturnal periodicity were proposed by Shostak and Esch (1990b). The first idea assumes that nocturnal emergence enhances the chance of being found by the next host in the life cycle. The second suggests that nocturnal release augments cercariae dispersal from the molluscan host. The third implies that diurnal periodicity reduces cercariae mortality. While these hypotheses were not directly tested, the Shostak and Esch (1990b) work suggested that nocturnal emergence could be advantageous if employed as a mechanism to avoid visual predators or if it increases the probability of encountering a nocturnally active second intermediate host. Regrettably, the questions revolving around adaptability remain unresolved.

Julie Williams (1989) came to work in my lab in the fall of 1987 from the University of Nebraska where she had come under the influence of Brent Nickol and John Janovy Jr. We had been working on various aspects of *Hal. occidualis* and *Hel. anceps* biology for several years when Julie made her appearance. By that time, we had a pretty good feel for the general characteristics of snail and trematode interactions. However, there were still some features of the trematode's population and community biology that were still questions for us.

Based on earlier mark–release–recapture studies by Jackie Fernandez and Tim Goater, we knew that the distribution of *Hal. occidualis* in the pond was not random—that there appeared to be some "hotspots," even in a very small impoundment like Charlie's Pond. We also knew that there were seasonal changes in the population biology of the parasite in the pond. Julie devised an approach that would allow her to follow seasonal change in individual snails while simultaneously assessing the microgeographic distribution of the parasite in the pond.

Her research design was very simple. She selected a U-shaped cove on the northwest side of the pond as a focus for her work. The piece of the shoreline was 68 m in length. At 1 m intervals, numbered stakes were inserted into the pond's substratum approximately 15–20 cm from the water's edge. She collected snails out to a depth of 50 cm, referring to her technique as "snail Braille." The 68 m shoreline was then divided into six sites, each ranging from 7 to 19 m in length. The sites were selected according to dominant biotic, or substrata, conditions, for example, *Typha* sp. stands, substratum quality, and so on.

Each snail was brought to the lab, checked for shedding cercariae, measured, and marked with its own identification number. It was then returned to its point of capture in the pond. Rather than trying to sample each site every couple of weeks, she set up a special procedure for checking the various sites on a random basis over the course of the study. Using these protocols, she was able to follow not only the timing of parasite recruitment, loss, and replacement but snail growth patterns and distributions of infected and uninfected snails. The results generated were quite revealing.

Her collection of snails ran for seven consecutive months in 1988, during which she examined approximately 3500 snails. Of these, 20% were recaptured at least once, providing her with the opportunity of developing a reasonably good picture of what was happening individually within a large group of snails over time. *Halipegus occidualis* and *Megalodiscus temperatus*, both autogenic parasites of green frogs, were regularly found at all six collecting sites. *Tylodelphys* (*Diplostomulum*) *scheuringi* was found only at sites one and two, which were both dominated by stands of *Typha*. The second intermediate hosts for *T.* (*D.*) *scheuringi* are mosquitofish, *Gambusia affinis*, which are much more abundant in the *Typha* sp. stands than in sites with more open water because the emergent vegetation offers at least some cover for avoidance of predators, that is, pied-bill grebes.

Snails infected with *Hal. occidualis* and *M. temperatus* typically occur in water that is less than 30 cm in depth, which is where green frogs spend a great deal of time submerged as ambush predators and where frog defecation and release of parasite eggs are most likely to take place. *Echinostoma* sp. infects smaller snails than any of the other trematodes for which the data are sufficient enough for adequate comparisons. Most of the recruitment of this parasite occurs in August when a large proportion of the *Hel. anceps* population is still small and apparently more susceptible to infection.

One of the other findings by Julie was that growth of infected snails appeared to be inhibited to some degree. This is in contrast to data from Crews and Esch (1986) and Fernandez and Esch (1991b), which indicated that infected snails are larger than uninfected snails. Goater (1989) suggested that the size differential may have resulted from an "... increased life span of infected hosts, a tendency for parasites to infect large hosts, or an increased prevalence of infection that occurs simply because larger, older snails have had more time to acquire infections."

On arriving at Wake Forest in pursuit of a master's degree, Brian Keas became involved studying the effect of parasitism on the reproduction and growth of *Hel. anceps* in Charlie's Pond. Brian was an exceptional student, but I have always felt that I steered him into what should have been a dissertation level of research rather than that of a master's degree. From my perspective, it nonetheless turned out to be an extraordinary piece of work—the quality of which matched its complexity.

The study was designed to "... discern the cause behind apparent incongruities between laboratory and field studies on snail growth" (Keas, 1995). Since explanations of gigantism revolve around the reallocation of energy typically employed for "... snail reproduction to either parasite reproduction (cercariae) or to snail growth, snails were raised on two diets of different quality, one high, which provided excess energy and one low in which energy would be limiting" (Keas, 1995). As an alternative to gigantism, fecundity compensation has been suggested as a possible host response. Fecundity compensation has been described as a way of reducing fecundity loss after infection by increasing egg production before infection can castrate the snail host (Minchella and Lo Verde, 1981). Accordingly, Brian infected snails prior to reproductive maturity so that compensation did not occur unless the rate of maturation increased, as well as after maturation so that compensation could occur. The complicated protocols involved, all experimental designs, data results, and statistical analyses, are provided in Keas (1995) and Keas and Esch (1996).

Brian's working hypothesis at the outset was that the presence of *Hal. occidualis* reduced snail reproduction but increased growth of *Hel. anceps* by inducing allocation of energy normally used for reproduction to somatic growth. His experimental design included raising snails that were provided low- and high-quality diets using three infection conditions, that is, uninfected snails (UNI), snails that were infected prior to reproductive activity (IPR), and snails infected after the outset of reproductive activity (IAR). The results revealed that the only snails to increase in wet mass (gigantism) were those infected prior to reproduction on a high-quality diet. Snails infected after reproductive activity began were not different in size from UNI controls.

Snails raised on the high-quality diet were consistently larger regardless of infection status, probably because growth rates in snails fed the high-quality diets were much faster prior to the beginning of reproduction. Regardless of the timing of infection, these snails consistently exhibited significantly reduced egg production compared with uninfected snails. Moreover, cercariae production by snails fed on the high-quality diet was significantly higher than snails fed a low-quality diet. Another feature of his results offered a possible explanation for the loss of infection under natural conditions in the pond, that is, prevalence of infection at the end of the experiments was lowest in snails fed on a low-quality diet. A comparison of the laboratory experiments with growth and fecundity data from field studies suggests that *Hel. anceps* in the field are likely protein deprived. Keas (1995) concluded that "Gigantism does not appear to be adaptive for the snail since survival was not enhanced, or for the parasite since cercariae production was the same for snails exhibiting gigantism or not." He went on to "… stress the need for caution when examining the fecundity and growth rates of both infected and uninfected snails in the laboratory, especially when comparing these data to other field or laboratory studies."

As I mentioned earlier, I think I made a mistake with my advice for Brian. The complexity of his problem was enormous, but his design was inventive. Typically, a biology master's student who works in the field has between 10 and 12 months to use in the conduct of their research. It takes time at the beginning to determine the nature of the question to be asked and a while longer to assemble the needed equipment, chemicals, and so on. If live vertebrate animals are involved, then they must be acquired and dealt with according to the criteria established by Animal Care and Use Committees (at least in the United States) before conduct of the research can be initiated. Then the research can follow. All of this must be done usually while the graduate student is taking courses, teaching labs, attending seminars, and so on. Brian did an especially beautiful piece of work. The results generated were right on target in understanding the physiology and ecology of infection by *Hal. occidualis* in *Hel. anceps*. If it had been a dissertation, another two full years of research time would have provided much more insight into a really complicated host–parasite relationship—too bad it was not!

Earlier in this chapter was a description of how Tim Goater discovered the presence of *Hal. eccentricus* in Charlie's Pond. After identifying the newly colonized species, several students, including Scott Snyder, Eric Wetzel, and Derek Zelmer, focused their attention on it, in addition to some unique attempts at further defining parasite population biology of *Hal. occidualis* in *R. clamitans*. Rather than detailing their stories in this chapter, I decided to split a discussion of their work and create a second chapter on research in the pond, which is why we have Pond I and Pond II. The latter chapter will also describe some sophisticated modeling efforts attempted by the last graduate student to work exclusively in the pond, Nick Negovetich.

References

Begon, M., J.L. Harper, and C.R. Townsend. 1990. Ecology: Individuals, populations, and communities, 2nd edition. Blackwell Scientific Publications, Inc., Boston, MA, 1049 p.

Crews, A.E., and G.W. Esch. 1986. Seasonal dynamics of *Halipegus occidualis* (Trematoda: Hemiuridae) in *Helisoma anceps* and its impact on fecundity of the snail host. *Journal of Parasitology* **72**: 528–539.

Crews, A.E., and G.W. Esch. 1987. Histopathology of larval trematode infections in the freshwater pulmonate snail, *Helisoma anceps*. *Journal of Invertebrate Pathology* **49**: 76–82.

Dronen, N.O. 1977. Host-parasite population dynamics of *Haematoloechus coloradensis* Cort, 1915 (Digenea: Plagiorchiidae. *American Midland Naturalist* **99**: 330–349.

Esch, G.W. 2007. Parasites and infectious disease: Discovery by serendipity, and otherwise. Cambridge University Press, Cambridge, 356 p.

Fernandez, J., and G.W. Esch. 1991a. The component community structure of larval trematodes in the pulmonate snail *Helisoma anceps*. *Journal of Parasitology* **77**: 540–550.

Fernandez, J., and G.W. Esch. 1991b. Effect of parasitism on the growth rate of the pulmonate snail, *Helisoma anceps*. *Journal of Parasitology* **77**: 937–944.

Fernandez, J., T.M. Goater, and G.W. Esch. 1991. Population dynamics of *Chaetogaster limnaei* (Oligochaeta) as affected by a trematode parasite in *Helisoma anceps* (Gastropoda). *American Midland Naturalist* **124**: 195–205.

Goater, T.M. 1989. The morphology, life history, ecology, and genetics of larval trematodes in the pulmonate snail, Helisoma anceps. Ph.D. dissertation. Wake Forest University, Winston-Salem, North Carolina, 155 p.

Goater, T.M., C.L. Browne, and G.W. Esch. 1990. On the life history and functional morphology of *Halipegus occidualis* (Trematoda: Hemiuridae), with emphasis on the cystophorous stage. *Journal of Parasitology* **76**: 923–934.

Goater, T.M., C.E. Goater, and G.W. Esch. 2014. Parasitism: The diversity and ecology of animal parasites, 2nd edition. Cambridge University Press, Cambridge.

Keas, B.E. 1995. The role of diet and reproductive maturity on the growth of Helisoma anceps (Pulmonata) infected by Halipegus occidualis (Trematoda). M.S. thesis. Wake Forest University, Winston-Salem, North Carolina, 52 p.

Keas, B.E., and G.W. Esch. 1996. The role of diet and reproductive maturity on the growth rate of *Helisoma anceps* infected by *Halipegus occidualis* (Trematoda). *Journal of Parasitology* **83**: 96–104.

Krull, W.A. 1935. Studies on the life history of *Halipegus occidualis* Stafford, 1905. *American Midland Naturalist* **16**: 129–142.

Macy, R.W., W.A. Cook, and W.R. DeMott. 1960. Studies on the life cycle of *Halipegus occidualis* Stafford (Trematoda: Hemiuridae). *Northwest Scientist* **43**: 1–17.

Minchella, D.J., and P.T. LoVerde. 1981. A cost of increased early reproductive effort in the snail *Biomphalaria glabrata*. *American Naturalist* **118**: 876–881.

Mulvey, M., T.M. Goater, G.W. Esch, and A.E. Crews. 1987 Genotype frequency differences in *Halipegus occidualis*-infected and uninfected *Helisoma anceps*. *Journal of Parasitology* **73**: 757–761.

Shostak, A.W., and G.W. Esch. 1990a. Temperature effects on survival and encystment of cercariae of *Halipegus occidualis* (Trematoda). *International Journal for Parasitology* **20**: 95–99.

Shostak, A.W., and G.W. Esch. 1990b. Photocycle-dependent emergence by cercariae of *Halipegus occidualis* from *Helisoma anceps*, with special reference to cercarial emergence patterns of adaptations for transmission. *Journal of Parasitology* **76**: 790–795.

Sousa, W.P. 1990. Spatial scale and the process structuring a guild of larval trematode parasites. *In* Parasite communities: Patterns and processes, G.W. Esch, A.O. Bush, and J.M. Aho (eds.). Chapman and Hall, London, p. 41–68.

Williams, J.A. 1989. A mark-recapture study on the infra- and component community dynamics of larval trematodes in the snail host, Helisoma anceps (Menke). M.S. thesis. Wake Forest University, Winston-Salem, North Carolina, 91 p.

7 The Pond: Part II

Time is an illusion, age is not.

This quote is mine (2014)

Tim Goater's discovery of *Halipegus* (=*Hal.*) *eccentricus* in the pond was pure serendipity. He certainly was not looking for it when Peg Mulvey at the SREL called to tell him of her findings. At that point, he had two species of *Halipegus*, one that cycled through *Helisoma* (*Hel.*) *anceps* and a second that employed *Physa gyrina* as the first intermediate host. Both parasites ended up in green frogs, the former under the tongue and the latter in the Eustachian tubes. Wendell Krull had resolved the life cycle of *Halipegus occidualis* in 1935, while Lyle J. Thomas described the cycle for *Hal. eccentricus* in 1939. Krull's account was accurate, but Thomas' description was incorrect, as I will explain in just a bit. During the first several years of our research on *Hal. occidualis* in the pond, we had focused on a number of host–parasite interactions, mostly involving the snail and frog hosts. Subsequent to finding *Hal. eccentricus* in the pond, however, we decided that we should do basically the same thing with the new parasite so that we could compare the biology of the two sympatric parasites.

Following Tim's discovery of *Hal. eccentricus*, the first graduate student to conduct work on the parasite was Scott Snyder, who came to Wake Forest in 1989. On his arrival, he and I had several discussions regarding what research he should do but decided that the population and community ecology of larval trematodes in the snail, *P. gyrina*, would be most suitable for his master's research.

Before considering the ecological features of the trematode's biology, however, it is necessary to first compare several aspects of the behavior and reproductive activity for both snails, that is, *Hel. anceps* and *P. gyrina*. As described in the previous chapter, the life span of *Hel. anceps* in the pond is 1 year (way up north, it is 2 years).

Ecological Parasitology: Reflections on 50 Years of Research in Aquatic Ecosystems,
First Edition. Gerald W. Esch.
© 2016 John Wiley & Sons, Ltd. Published 2016 by John Wiley & Sons, Ltd.

Each annual cohort of snails begins to die in late June and continues into July. Seasonally, reproduction by mature *Hel. anceps* begins in March when water temperatures climb above 15°C. Reproduction declines in midsummer because the older snails are dying and many of the new snails are still juveniles. However, reproduction climbs to a second peak toward the end of July, steadily declines into the fall months, and stops completely by mid-September as water temperatures begin to fall. In contrast to *Hel. anceps*, the life span of *P. gyrina* is 3–4 months during the summer. After reaching sexual maturity, these snails produce eggs as long as they remain alive and uninfected. Relative to *Hel. anceps*, *P. gyrina* exhibits substantially greater vagility in the pond, which explains why our attempts to conduct mark–release–recapture studies on the latter species were unsuccessful. Over a 14-month period, 97% of marked *Hel. anceps* were recovered from within 1 m of their release point. In contrast, 28% of marked *P. gyrina* had moved one and a half meters after just 2 weeks following release.

In 1984, the component communities of larval trematodes in *Hel. anceps* included seven species. Scott initially identified six species in *P. gyrina*. In both snails, a different species of *Halipegus* was dominant, that is, *Hal. occidualis* in *Hel. anceps* (overall prevalence ~60%) and *Hal. eccentricus* in *P. gyrina* (overall prevalence ~49%). During the 1980s and 1990s, the prevalences of both *Halipegus* species significantly varied seasonally. The numbers of *Hel. anceps* infected with *Hal. occidualis* were initially very high but tapered off dramatically after 2000. The prevalences of *Hal. eccentricus* were always less than *Hal. occidualis*.

Even though there are some similarities in the parasite component communities, there are also several noteworthy differences as well. From 1983, when the first collection of snails was made in Charlie's Pond, through 2006, approximately 18,000 *Hel. anceps* were collected and checked for trematode infections. Among all of these snails, multiple infections were seen less than 20 times! As part of Scott Snyder's master's research (Snyder and Esch, 1993), he examined approximately 1200 *P. gyrina* from April 1991 to March 1992; 33.9% were infected, and of these, 18.4% were double, or triple, infections. The most frequent combination was *Hal. eccentricus* and *Haematoloechus* (=*Hae.*) *complexus* (both AU=autogenic); the latter is a lung fluke in several *Rana* spp. The next most common pair was *Hae. complexus* and *Glypthelmins quieta* (also both AU), the latter a gut fluke in several *Rana* spp. These associations are of interest for a couple of reasons. First, as noted, all four of these parasites are AU, and moreover, all four employ *Rana clamitans* as the definitive host. The most common AL species infecting *P. gyrina* was *Echinostoma trivolvis*, one of the erratic, or "come and go," trematodes. In 1984, *E. trivolvis* was not present in *Hel. anceps*; the dominant AL species in that year were *Tylodelphys* (*D.*) *scheuringi* and *Zygocotyle lunata*, at about 3%. Five years later, in 1989–1990, *Haematoloechus* (=*Hae.*) *longiplexus* (AU), *Megalodiscus temperatus* (AU), *Petasiger nitidus* (AL), *E. trivolvis* (AL), and *Plagitura parva* (AU) joined the *Hel. anceps* parasite component community, replacing *Z. lunata* (AL), *G. quieta* (AU), *Spirorchis* sp. (AL), and

Clinostomum sp. (AL). *Tylodelphys* (*D.*) *scheuringi* was present again, with its prevalence slightly higher than 16%.

The differences in infracommunity structure within *Hel. anceps* and *P. gyrina* have been attributed to several factors. For *Hel. anceps*, the absence of multiple infections is, in part, due to low vagility, that is, the less distance traveled, the less exposure to both sessile (eggs) and motile (miracidia) infective stages. However, during Jackie Fernandez's work in the pond, she also was able to ascertain the presence of a dominance hierarchy among several species of trematodes infecting *Hel. anceps*. At the top of the hierarchy were parasites with intramolluscan rediae stages. While *Hal. occidualis* was high in the hierarchy, the top competitor was *E. trivolvis* whose rediae are considerably larger than those of *Hal. occidualis*. In contrast, Scott determined that a dominance hierarchy did not exist in *P. gyrina*. He suggested that this trait, along with the snail's greater vagility, could account for so many multiple infections in single hosts.

Another feature contributing to the different infracommunity dynamics in the two snail species rests with differences in their life histories. The *Hel. anceps* population turns over in midsummer, which causes a total reset of the compound community structure each year, while egg production and steady recruitment of *P. gyrina* assures the constant addition of new "patches" for parasite recruitment throughout the summer months. Another characteristic that may impact the infrastructure difference between *Hel. anceps* and *P. gyrina* in Charlie's Pond is the fact that the *Hel. anceps* population turns over in June/July, while *P. gyrina* is apparently turning over constantly, with the life span of the species lasting for about 3–4 months. It would be likely to see the echinostomes in spring and fall, corresponding with waterfowl migration patterns, and *Glypthelmins* sp. (a gut parasite in frogs) present in summer only.

Another master's student, Kelli Sapp, came from Methodist College over in Raleigh, North Carolina. My guess is that Herman Eure had a huge impact on this decision, because after completing her master's degree here and then her PhD with Sam Loker at the University of New Mexico, she and Herman were married in Davis Chapel here on our campus in Winston–Salem. I am very proud to say that I was Herman's "best man" at the wedding and am pleased to note that I drove them to their home for the wedding reception immediately after they were married—yes, in my convertible and with the top down (the "dean machine"—so-called by my graduate students since I was dean of the graduate school here at the time)!

Kelli's research focused on an examination of temporal and spatial factors that influence the structure of infracommunities in both *Hel. anceps* and *P. gyrina*. We found that some of Kelli's work would dovetail nicely with some of the spatial and distributional efforts of a later graduate student, Derek Zelmer. Jackie Fernandez had established that intramolluscan trematode antagonism played at least some sort of role in structuring infracommunities in *Hel. anceps*—also recall that infracommunities in *Hel. anceps* are rather depauperate. In contrast, however, Scott Snyder determined that double and even triple infections were relatively common in the sympatrically distributed *P. gyrina* in Charlie's Pond. Moreover, there was substantial variability in distribution of parasites in both snail species at three different collecting sites in the pond.

Kelli also found that the species in the infracommunities of *Hel. anceps* and *P. gyrina* could be placed into two distinct clusters, that is, those that most frequently overlapped spatially and those that did not (Sapp and Esch, 1994). In *Hel. anceps*, the overlapping group included *Hal. occidualis*, *Hae. longiplexus*, and *M. temperatus*; in *P. gyrina*, the group included *Hal. eccentricus*, *Hae. complexus*, *M. temperatus*, and *G. quieta*. There are several interesting features regarding these distributions. First, in both snail species, sympatry occurred in the largest snails rather than smaller ones. Second, the overlapping parasites mostly occurred in hosts that occupy shallow water and closer to shore. Third, all of these trematodes are AU. Finally, five of the trematode species in these two groups infect the snail intermediate host via ingestion of eggs. Since green frogs are ambush predators and spend a great deal of their time in shallow water near shore, it is reasonable to expect that frog defecation, trematode egg release, and accidental ingestion of eggs by the snails as they graze collectively account for the spatial aggregation of these species in *Hel. anceps* and *P. gyrina*. In contrast to the other trematodes in the two groups, *M. temperatus* infects both species of snails via free-swimming miracidia. However, we would expect *M. temperatus* to cluster with the other ranid-infecting trematodes since their eggs would be released from *R. clamitans* and hatch in the same localities occupied by the other flukes.

Eric Wetzel graduated from Millersville University in Pennsylvania and came to Wake Forest where he worked with Dr. Pete Weigl on the biology of a rhabditid nematode, *Strongyloides robustus*, in flying squirrels. After earning his master's degree with Pete, Eric decided he would switch laboratories and pursue his PhD with me. We discussed his research goals and determined that *Hal. eccentricus* offered a really good opportunity for his dissertation research. There were a couple of features on which he really wanted to concentrate. One had to do with the confirmation of Thomas' (1939) life cycle description, and the second was to examine the parasite's population biology in the green frog from Charlie's Pond. The baseline question was, "How similar are *Hal. occidualis* and *Hal. eccentricus* biologically?" Even though they occupied the same definitive host, they were not sympatric when it came to their sites of infection in *R. clamitans*, that is, *Hal. occidualis* is found under the tongue and *Hal. eccentricus* in the Eustachian tubes.

The first thing that Eric did was to collect more than 100 *R. clamitans* tadpoles from Charlie's Pond and necropsy them, but none was infected with *Hal. eccentricus*. If Lyle Thomas (1939) had been correct in his proposed life cycle for *Hal. eccentricus*, the parasite should have been present in at least some of the tadpoles, but it was not. During his work on the population biology of *Hal. eccentricus* in adult definitive hosts, Wetzel (1995) unexpectedly, and consistently, found immature specimens of *Hal. eccentricus*. This observation, coupled with the fact that some of the green frogs he captured were at least 3 years old, clearly indicated that the parasite did not use green frog tadpoles as a mode of infection, at least in Charlie's Pond. Interestingly, he also found a significant positive correlation between the population biology of the two species of *Halipegus* in green frogs, except that the peak prevalence of *Hal. eccentricus* occurred in June, while *Hal. occidualis* hit the highest level in July. At this point, Eric began to think that Thomas' three-host life cycle description for *Hal. eccentricus* was incorrect. Similarities between population biology of the two *Halipegus* species were leading Eric to believe their life cycles were essentially the same, that is, four hosts for both.

Over a 3-year collecting period in Charlie's Pond, Eric caught, marked, and released approximately 150 *R. clamitans*. Comparisons regarding the population biology of the two *Halipegus* species are of interest. For example, recruitment of the two species always began at the same time, in April, which coincides with the initial emergence of dragonflies in the pond. Prevalences and intensities of *Hal. eccentricus* were always lower than for *Hal. occidualis*. Eric suggested that the explanation for the difference, at least in part, rested with a much smaller amount of surface area available for attachment by *Hal. eccentricus* in the Eustachian tubes as compared with under the tongue for *Hal. occidualis*. The presence of adult Halipegus spp. in April also suggests that both species overwintered in the frogs, thereby confirming his laboratory observations.

Eric also had an opportunity to compare his collection data with those compiled by Tim Goater several years previously and with those of Derek Zelmer who collected data a few years after Eric. This comparison is of interest because the field data clearly point to a decline in the frog population size over the years. These declines were also accompanied by reductions in the infrapopulation sizes of both *Halipegus* species.

Another feature of the population biology of Halipegus spp., in particular *Hal. occidualis*, was the way in which some infrapopulations increased and decreased over time. The maximum size of an *Hal. eccentricus* infrapopulation in the Eustachian tubes was 12, with most having just five or six worms. In contrast, Eric found one frog with 46 *Hal. occidualis* under the tongue and several with 30. Similar intensity data had been previously observed by Goater (1989). Tim opined that large variance/mean ratios (VMR) in frogs were due to the chance recruitment of heavily infected odonates. Although Tim did not collect dispersion data in odonates, Eric did (Wetzel and Esch, 1996). He found that the frequency distributions of *Hal. occidualis* infrapopulations were overdispersed in every species of odonate examined from the pond and proposed that the very sharp (rapid) declines in *Hal. occidualis* infrapopulation sizes over very brief time periods were due to density-dependent expulsion of worms by the host. For example, he identified a marked frog that had 25 adult worms initially, but at recapture a week later, the entire infrapopulation was gone. In hosts with heavy infections, there was typically severe inflammation in the tissues under the tongue at the *Hal. occidualis* attachment sites. In a recent conversation I had with Tim, he reminded me that under the tongue of green frogs, there is a pair of prominent blood vessels just below the tissue surface and that adult *Hal. occidualis* always attach to the tissue immediately above these blood vessels. Intensity of inflammation is much more likely when these helminth parasites are "bunched," rather than scattered or fewer in number.

Derek Zelmer was the third of my Canuck graduate students. He came in the fall of 1994, having completed his undergraduate and master's degrees at the University of Calgary. Derek's dissertation research was wide in scope. It took him from describing the mechanism for release of the cercaria body from the modified tail to testing a sophisticated mathematical model dealing with the treatment of human helminth parasites using the *Hal. occidualis*/*R. clamitans* system.

In the previous chapter, I briefly described the explosive release of the cercariae bodies from within the tails of cystophorous cercariae. I have searched my memory, but I can think of only one other biological "explosion" of this sort, and that is associated with the manner in which cnidarian nematocysts release a coiled tubule that delivers a toxin into unsuspecting prey organisms (I was reminded by Derek that polar filaments of myxozoans exhibit a similar phenomenon—although myxozoans are now considered as cnidarians). The trigger mechanism for the nematocyst is associated with the release of stored calcium that creates a very large concentration gradient across the cell's plasma membrane. This is followed by the rapid entry of water, creating osmotic pressure that forces the coiled tube to explode from inside the nematocyst (Nuchter et al., 2006). The emergence speed is estimated at about 700 ns, with a force of about 40,000 g's. The process of releasing the cercaria body is obviously similar to the nematocyst explosiveness, but Derek did not pursue the mechanism for discharge by *Hal. occidualis* any further.

Throughout the research history of *Hal. occidualis*, there has been considerable discussion regarding the role of odonates (dragonflies) in the life history of the parasite. Krull (1935) successfully resolved the life cycle, but he was unsure as to whether odonate naiads are obligate intermediate hosts or not. Macy et al. (1960) concluded that a naiad was a paratenic host, but Goater et al. (1990) countered that it was an obligate intermediate host. Finally, Zelmer (1998) provided clear evidence that the naiads are, in fact, paratenic hosts. The problem was easily resolved. Derek discovered that if ostracods were exposed to cystophorous cercariae in the lab and maintained for at least 4 weeks and then fed directly to green frogs, adult *Hal. occidualis* would develop under the tongue. Very simply, this indicates that odonate naiads are paratenic hosts and are required insomuch as they bridge a trophic gap in order to continue the parasite's life cycle. However, there is no evidence to suggest that these microcrustaceans are included in the diet of adult frogs. Moreover, we seriously doubt frogs can see them.

A German geographer, Carl Troll, coined the term landscape ecology in 1939. By 1980, the concept had developed into a significant subdiscipline of ecology.

The Russian epidemiologist Pavlovsky (1966) introduced the idea of focal nidality, or focality, and employed the term landscape epidemiology for the first time. Since then, with the use of remote sensing and Geographic Information System (GIS), this relatively new idea has been used to identify localities where specific disease transmission is a high probability. Working with Eric Wetzel, Derek applied this concept to Charlie's Pond and the transmission of *Hal. occidualis*. They were able to identify four "hotspots" in the pond where they found maximum transmission of the parasite was occurring and to recognize which physical and biologic characteristics these localities had in common. Plotting parasite infrapopulation size against the location of host capture identified the most active transmission sites. All four of the "hotspots" were characterized by shallow water and emergent vegetation. The green frog is an ambush predator that favors shallow water in which to wait. Emergent vegetation also offers camouflage cover for the frogs and a way for odonate naiads to eclose above the water surface. The vegetation also serves as a "trap" for the accumulation of leaf litter and helps to stimulate the development of aufwuchs communities for consumption by the nearly sessile *Hel. anceps* snails. In other words, most of the physical and biological requirements for the parasite to successfully complete its life cycle are present in these so-called hotspots.

One of the unassailable characteristics of population biology associated with adult helminth parasites is the occurrence of overdispersed frequency distributions. As previously described, it was proposed by Crofton (1971a, 1971b) that an aggregated (overdispersed) parasite distribution should "… have a stabilizing influence on parasite populations through regulation of host populations mediated by death of heavily infected hosts (Anderson and May, 1978), or through concentration of density-dependent effects on infrapopulations (Shaw and Dobson, 1995)" (Zelmer, 1998). This concept has been employed, vis-à-vis, through the use of mathematical modeling, in the development of treatment protocols that focus holistically on human populations rather than on individuals. The idea in this case is to concentrate on reduction of morbidity rather than prevalence of disease produced by parasitic helminths (Anderson and May, 1982, 1985; Anderson and Medley, 1985).

The treatment process in human populations can be separated into three schemes. A mass treatment protocol requires therapy for all members of a population, regardless of known epidemiological characteristics for disease transmission. Selective treatment would focus only on those members of a population with the heaviest infections; therefore, because of overdispersion, only a relatively small number of hosts would be treated. A third option, targeted treatment, would focus on a part of the population in which prior epidemiological investigation would predict the segment of the population that should carry the most parasites. Particular behaviors, cultural practices, genetic predisposition, and so on increase infection levels in these individuals to a higher level than the general population, that is, individuals with a higher risk are treated.

While the idea for testing these treatment protocols is intriguing, all of the schemes are experimentally problematic because natural host–parasite systems require the quantification of parasites to be made before and after treatment in order to assess its effectiveness. However, this handicap does not apply for the *Hal. occidualis/R. clamitans* system. Frogs can be easily captured and marked for individual identification, and *Hal. occidualis* infrapopulations can be easily counted, without treatment or without having to kill the host and then count the parasites at necropsy. Since *Hal. occidualis* adults can be enumerated without killing the host, the frog/*Hal. occidualis* model also can be followed longitudinally.

It was after a conversation with Celia Holland at Trinity College in Dublin that I suggested to Derek Zelmer he use this model in his dissertation research. His idea was to "… test the effect of selective treatment on the stability of a well-defined host-parasite system that is conducive to the precise enumeration of worm burden densities, thereby more clearly assessing the role of aggregation in promoting parasite component population stability" (Zelmer, 1998). The experimental design he decided on using included the elimination of approximately 50% of the highly overdispersed adult worm population, with one worm to be left in each selectively "treated" frog so that the effects of decreased worm prevalence would not compound the effects of reduced abundance or worm aggregation.

Infrapopulations of *Hal. occidualis* were tracked over a 3-year period, beginning in 1995, with the establishment of baseline estimates for infrapopulation sizes. Then, in 1996, all but one worm was removed from heavily infected frogs (those with >15 worms). This procedure reduced the estimated component worm population by 45% and the mean intensity and VMR of *Hal. occidualis* in the pond by 85% and 63%, respectively. The next year (1997), collection data indicated a return to mean intensity, aggregation, and recruitment rates virtually identical to those of 1995, the baseline year. These results indicate no effect of worm removal and high stability for this host–parasite system. Based on these observations, he concluded that transmission in his system is controlled by the prevalence of adult worms, not by their intensity of infection (Zelmer, 1998). Following an analysis of field studies using the three treatment procedures up to that point in time, he asserted that "… treatment protocols aimed at disrupting helminth transmission based on theoretical models might benefit from a re-evaluation that considers distribution of infective stages as the result of infection prevalence."

Derek made yet another significant observation during his dissertation research. As was noted previously, the colonization of the pond by *P. gyrina* brought several new species of trematode into the parasite mix in Charlie's Pond. Interestingly, the component community of *P. gyrina* was dominated by *Hal. eccentricus*, a congener of *Hal. occidualis* in *Hel. anceps*. Prevalence of *Hal. eccentricus* in *P. gyrina* at first was quite high, almost 50% (Snyder, 1992). Sapp (1993) observed a slight decline to 30% at several sites in the pond. By 1995, the prevalence had dropped to less than 1%, and Zelmer (1998) found the parasite in just two frogs during a 2-year period of sampling that included 180 green frogs.

Anna Schotthoefer graduated from Cornell University in 1993 with a BS degree and came to work in my lab in 1996. In the acknowledgments of her master's thesis, she credits Gene Burreson at the Virginia Institute of Marine Science for steering her into parasite ecology. Therefore, I too am grateful to Gene because Anna provided a significant contribution to our work out in Charlie's Pond. One of Anna's thesis objectives was to "... verify and examine the potential effects of the abrupt decline in *Hal. eccentricus* on the larval trematode component community of *P. gyrina*."

In her study, she found six species of trematodes in *P. gyrina*. Three of these species employ *R. clamitans* as definitive host. In approximately 2400 snails, the prevalence of *Hal. eccentricus* was less than 1%, confirming the very sharp decline observed by Derek Zelmer and Eric Wetzel. None of the snails carried *Hae. complexus*, an autogenic species that exhibited a prevalence of 19.5% just 3 years prior (Snyder and Esch, 1993). It is of interest that *Hae. complexus* uses *R. clamitans* as a definitive host, along with *Hal. eccentricus*. Derek had suggested in 1995 that the green frog population in Charlie's Pond was in decline based on frog capture and recapture data. However, I would suggest caution here since *M. temperatus*, which also uses green frogs as its definitive host, peaked at approximately 2% of the *P. gyrina* and less than 1% of the *Hel. anceps* (Schotthoefer, 1998). We have made no effort to collect frogs for several years, but Collin Russell (personal communication), an undergraduate student who worked in the pond in 2013, indicated that tadpoles "... were all over the place."

As was described in Chapter 6, Charlie's Pond and the two *Halipegus* species were the research focus for a long string of graduate students beginning in 1984. The last of these students was Nick Negovetich, who completed his master's thesis in 2003 and his PhD in 2007. Nick was an undergraduate student of Eric Wetzel at Wabash College who decided that an academic career was a great way to spend his life and that parasitology should be his focus.

Each student with whom I have been in contact has had a unique set of qualities. In Nick's case, one of his strongest assets was his background in mathematics and statistics, which he brought to bare on his PhD research. When Nick began his dissertation work, I think that both he and I believed that he would be the last graduate student to work in Charlie's Pond. Accordingly, we wanted him to do something special with his dataset, and he did. When Nick began, he identified several objectives that he wanted to achieve. First, how had the trematode component community changed in the pond from 1984 through 2006? Second, what is the life history cost of trematode infection for *Hel. anceps* in Charlie's Pond? Third, he wanted to quantitatively estimate the cost of parasitic castration for *Hel. anceps* in the pond using a matrix population model.

By the time Amy Crews completed her research in the spring of 1986, the component community consisted of seven species of trematodes, two of which were autogenic (AU) and the rest were allogenic (AL). Nick believed that Amy's diagnosis of *G. quieta* was incorrect and that one of the misidentified parasites was actually *Hae. longiplexus* (AU). He also asserted that *Z. lunata* (AU) identified in 1984 was more likely *M. temperatus* (AU). In 2005–2006, Nick identified 13 species in *Hel. anceps* from the pond. The new ones included *P. parva* (AU), which uses common newts as definitive hosts; two unknown spirorchiids (both AL); *Cercaria marginata*; *Uvulifer ambloplitis* (AL); and an unknown armatae cercaria. Three of these species had been previously reported in the pond and two were new. Overall, five species were AU and eight were AL. The most consistent species from year to year were AU, as might be expected.

Nick employed various reproduction experiments and mark–recapture methods to "... directly measure key life history traits of *Hel. anceps* in Charlie's Pond" (Negovetich, 2007). Infection by *Hel. anceps* resulted in complete castration of the snails, although he confirmed Goater (1989) that approximately 25% of the castrated individuals lost their infections while overwintering and that many resumed reproductive activity the following spring. He also determined that neither survival nor growth rates were different between infected and uninfected snails, reproducing observations made before by Fernandez and Esch (1991). As was previously noted (Fernandez and Esch, 1991), Negovetich (2007) observed that there was seasonal variation in parasite recruitment that did not vary with snail size. He suggested that the correlation between host size and prevalence of *Hal. occidualis* was not related to differential mortality or changes in growth rates, but to rapid growth rates of smaller infected snails.

The most significant contribution by Nick came with the development of the first mathematical model to describe the dynamics of a snail population. He was able to construct a period matrix product model using estimates of fecundity, survival, growth rates, and infection probabilities with data generated by his own mark–recapture studies as well as those generated by previous work, extending back to the results of Crews (1985) who conducted the first research in Charlie's Pond.

Another important aspect of the study by Negovetich (2007) was his observation regarding the effect of castration on the life history of *Hel. anceps* in Charlie's Pond. When work in the pond began in 1984, snails smaller than 8.0 mm did not reproduce. Goater (1989) supported Crews' observation by stating that *Hel. anceps* reaches sexual maturity at 7.5–8.0 mm. In his master's thesis in 2003, Nick stated that the smallest snail to oviposit in 2002 was 5.65 mm. In 2005 and 2006, snails that were 6.0 mm in size were laying eggs. He concluded that the size of ovipositing snails had significantly declined since research began in 1984, suggesting that castration by *Hal. occidualis* had driven this change in the *Hel. anceps* life history strategy in Charlie's pond. Nick noted that most of the reports from the 1980s and

1990s indicated that prevalences of *Hal. occidualis* in *Hel. anceps* were greater than 50%. Negovetich and Esch (2007, p. 1315) stated that

> With 50% of the population reproductively dead, the individuals that were most fit are those that reproduced prior to castration. Smaller snails are less likely to be infected than larger snails, especially in a system where individuals are continuously exposed, or nearly so, to the infective stages of castrating trematodes. Over time, the population of *Hel. anceps* should evolve toward reproduction at smaller sizes.

Apparently, this is what has happened in Charlie's Pond. It is an excellent example of natural selection, in the classic sense!

Most recently, I have had three undergraduate students (Collin Russell, Courtney Sump, and Tina Casson) working in the pond, collecting both *P. gyrina* and *Hel. anceps* from April into October in 2012–2014. In 2012, a total of six species of trematodes were identified in *Hel. anceps*, and two in *P. gyrina*. In 2013, only one trematode species was being shed from *P. gyrina* and just two species from *Hel. anceps*. In 2014, there were again two species in *Hel. anceps* and four in *P. gyrina*.

An obvious question is, what happened to the parasites? An addendum (Chapter 13) to the present chapter and Chapter 6 summarizes the work accomplished by the three undergraduate and three graduate students with close ties to the most recent work in Charlie's Pond (Russell et al., 2015). The addendum offers several suggestions as to the changing situation regarding the snails and their trematode communities in Charlie's Pond. From 1984 to the present time, the diversity of trematodes in the pond has perceptibly changed, and we think we have a reasonable hypothesis to explain the situation as it now stands.

In closing the present chapter, I would strongly encourage any young parasitologist who is just beginning a career to find something like Charlie's Pond, start a sampling program, and keep it going for a long, continuous stretch of time. My students and I promise that it will pay great dividends!

References

Anderson, R.M., and R.M. May. 1978. Regulation and stability of host-parasite population interactions. I. Regulatory processes. *Journal of Animal Ecology* **47**: 219–247.

Anderson, R.M., and R.M. May. 1982. Population dynamics of human helminth infections: Control by chemotherapy. *Nature* **297**: 557–563.

Anderson, R.M., and R.M. May. 1985. Helminth infections of humans: Mathematical models, population dynamics and control. *Advances in Parasitology* **24**: 1–101.

Anderson, R.M., and G.F. Medley. 1985. Community control of helminth infections by mass and selective chemotherapy. *Parasitology* **90**: 629–660.

Crews, A.E. 1985. Seasonal dynamics of larval trematode infections in Helisoma anceps and the effect of trematode infection on the fecundity of the host snail. Master's thesis. Wake Forest University, Winston-Salem, North Carolina, 80 p.

Crofton, H.D. 1971a. A quantitative approach to parasitism. *Parasitology* **62**: 179–193.

Crofton, H.D. 1971b. A model of host-parasite relationships. *Parasitology* **63**: 343–364.

Fernandez, J., and G.W. Esch. 1991. The component community structure of larval trematodes in the pulmonate snail *Helisoma anceps. Journal of Parasitology* **77**: 540–550.

Goater, T.M. 1989. The morphology, life history, and genetics of Halipegus occidualis (Trematoda: Hemiuridae) in molluscan and amphibian hosts. Ph.D. dissertation. Wake Forest University, Winston-Salem, North Carolina, 155 p.

Goater, T.M., C.L. Browne, and G.W. Esch. 1990. On the life history and functional morphology of *Halipegus occidualis* (Trematoda: Hemuridae), with emphasis on the cystophorous stage. *Journal of Parasitology* **76**: 923–934.

Krull, W.H. 1935. Studies on the life history of *Halipegus occidualis. American Midland Naturalist* **16**: 129–143.

Macy, R.W., W.A. Cook, and W.R. DeMott. 1960. Studies on the life cycle of *Halipegus occidualis* Stafford, 1905. *Northwest Science* **34**: 1–17.

Negovetich, N.J. 2007. Trematode communities in Charlie's Pond: The individual and population cost of infection in the pulmonate snail Helisoma anceps. Ph.D. dissertation. Wake Forest University, Winston-Salem, North Carolina, 173 p.

Negovetich, N.J., and G.W. Esch. 2007. Long-term analysis of Charlie's Pond: Fecundity and trematode communities of *Helisoma anceps. Journal of Parasitology* **93**: 1311–1318.

Nuchter, T., M. Benoit, U. Engel, S. Ozbek, and T.W. Holstein. 2006. Nano-second scale kinetics of nematocyst discharge. *Current Biology* **16**: R316–R318.

Pavlovsky, E.N. 1966. Natural nidality of transmissible diseases. University of Illinois Press, Urbana, IL, 261 p.

Russell, C., T. Casson, C. Sump, K. Luth, M. Zimmermann, N. Negovetich, and G. Esch. 2015. The catastrophic collapse of the larval trematode component community in Charlie's Pond (North Carolina). *Journal of Parasitology* **101**: 116–120.

Sapp, K.K. 1993. Effects of microhabitat on the transmission dynamics of larval trematodes. Master's thesis. Wake Forest University, Winston-Salem, North Carolina, 60 p.

Sapp, K.K., and G.W. Esch. 1994. The effects of spatial and temporal heterogeneity as structuring forces for parasite communities in *Helisoma anceps* and *Physa gyrina. American Midland Naturalist* **132**: 91–103.

Schotthoefer, A.M. 1998. Spatial variation in trematode infections and fluctuations in component community composition over the long-term in the snails Physa gyrina and Helisoma anceps. Master's thesis. Wake Forest University, Winston-Salem, North Carolina, 72 p.

Shaw, D.J., and A.P. Dobson. 1995. Patterns of macroparasite abundance and aggregation in wildlife populations: A quantitative review. *Parasitology* **111**: S111–S133.

Snyder, S.D. 1992. Trematode community structure in the pulmonate snail Physa gyrina and the effect of parasitism on fecundity of the snail host. Master's thesis. Wake Forest University, Winston-Salem North Carolina, 77 p.

Snyder, S.D., and G.W. Esch. 1993. Trematode community structure in the pulmonate snail, *Physa gyrina. Journal of Parasitology* **79**: 205–215.

Thomas, L.J. 1939. Life cycle of a fluke, *Halipegus eccentricus*, found in the ears of frogs. *Journal of Parasitology* **25**: 207–221.

Wetzel, E.J. 1995. Seasonal recruitment and infection dynamics of Halipegus occidualis and Halipegus eccentricus (Digenea: Hemiuridae) in their arthropod and amphibian hosts. Ph.D. dissertation. Wake Forest University, Winston-Salem, North Carolina, 120 p.

Wetzel, E.J., and G.W. Esch. 1996. Seasonal population dynamics of *Halipegus occidualis* and *Halipegus eccentricus* (Trematoda: Hemiuridae) in their amphibian host, *Rana clamitans*. *Journal of Parasitology* **82**: 414–422.

Zelmer, D.A. 1998. Life history and transmission dynamics of Halipegus occidualis (Digenea: Hemiuridae). Ph.D. dissertation. Wake Forest University, Winston-Salem, North Carolina, 164 p.

8 The Big Lake

Art is long, life short; judgment difficult, opportunity transient.
Wilhelm Meister's Apprenticeship, Johann von Goethe
(1749–1832)

After 1972, and throughout the remaining 1970s decade, my primary bases of operations for research and those of my students were Gull Lake in Michigan and the Par Pond reservoir at the Savannah River Plant (SRP) in South Carolina. Toward the end of the 1970s, I began to feel that I should move my research operations closer to home. For nearly 3 years, I had watched Joe Camp drive down to the Savannah River Ecology Laboratory (SREL) from Winston-Salem on a regular basis so that he could collect mosquitofish, and it was a pretty good drive, about 3 hours each way. I always felt bad about this arrangement, especially when I discovered that one of the larval trematodes on which he was working was a regular visitor to Charlie's Pond.

Ron Dimock, a colleague of mine in the Biology Department at Wake Forest, and I had tried to get something started locally in the early 1970s, but the effort was not very successful. In 1973, he and I convinced Duke Energy to fund a study on *Proteocephalus ambloplitis* in a new reservoir, Belews Lake (Figure 8.1). They had constructed it to supply water for a new coal-fired power plant not far from Winston-Salem. Unfortunately, our research was largely unsuccessful, at least from the standpoint of the bass tapeworm (in fact, our gill nets were stolen the first night we set them!). However, we did make some extensive copepod collections over a 2-year period, which were to come in handy about 10 years later.

In 1978, Joe Camp inquired if I would be interested in taking on a graduate student who had just finished up with Harry Huizinga, an old friend who was to eventually help Terry Hazen and me do some histopathology on largemouth bass

Ecological Parasitology: Reflections on 50 Years of Research in Aquatic Ecosystems,
First Edition. Gerald W. Esch.

Figure 8.1   A group of students take a look at Belews Lake. Charlie's Pond is located behind them, about 20 m away. Goater et al. (2013). Reproduced with permission of Cambridge University Press.

during our work on red sore disease down at SREL. I replied, sure, have him apply—the student's name was Willard (Bill) O. Granath, and he accepted our offer of a teaching assistantship.

In the late 1970s, a very unusual phenomenon was to take place out at Belews Lake, one that was to open wide the gate for some very unique field studies for Bill, also Dave Marcogliese and Mike Riggs, both of whom were to become my PhD students and somewhat overlap with each other and with Bill Granath. Because the power plant at Belews Lake is coal fired, a considerable amount of fly ash is produced on a daily basis. During the early years of operations at Belews Lake, fly ash was converted into a slurry and transported to a settling basin about a quarter mile from the reservoir. The idea was to allow particulates to settle and then to return the "clean" water back to the reservoir.

I have heard some folks declare that "The road to hell is paved with good intentions." I can honestly say that Duke Energy had good intentions in disposing of the fly ash; in fact, I would assert that during the early years in Belews Lake, they had one of the best environmental records for an energy company in the United States. However, what they did not know was that the coal they were burning was very high in selenium (an element that was to haunt them from that time period to the present day). Normally, selenium is an antioxidant that is a requisite in very small amounts by all animals. However, in large quantities, it will cause sterility in exposed animals, and in very large quantities, it is lethal. During the first few years of plant operations

(~1970–1971), the reservoir possessed 24 piscine species. However, by the end of the 1970s, the number of fish species sharply declined from 24 to just 4, that is, leaving behind a few carp and channel catfish and a very large number of fathead minnows (which were actually introduced early during the environmental "trauma" induced by selenium pollution) and mosquitofish. The latter two fish species were to become dominant (for a while at least) in the main body of the lake. The decline in extant fish species was attributed to exceedingly high concentrations of selenium that had accumulated in the main body of the lake during the first several years of Duke Energy's operations.

Another interesting event had also occurred at the south end of the reservoir, near the headwaters of the lake. A chemocline had developed. This phenomenon is analogous to a thermocline, which, as described in Chapter 3, involves the creation of a thermal and density gradient that will horizontally separate the epilimnion and hypolimnion in a stratified lake. A chemocline does essentially the same thing, except in this case it vertically separated the selenium-polluted main body of the lake from the unpolluted headwaters of the reservoir. Accordingly, even though the piscine fauna in the main body of the lake had been decimated, the headwaters retained all species of the original fish community.

Soon after Bill Granath arrived at Wake Forest in the fall of 1979, he came to my office and announced that he was going out to Belews to do some seining for fish. I responded by telling him to be careful because there is very little littoral zone in the lake (you could go out 15 ft from the shoreline and find yourself in 15 ft of water in most places). I also recall telling him, "Bill, you are not going to find anything— there are just two species of fish that are abundant and those are very small. They are not going to have any parasites worth the time and effort of a dissertation!"

Bill went anyway. Later that afternoon, he came by my office and invited me to the lab. He asked me to take a look into his dissecting scope. There, right before my unbelieving eyes, was a tapeworm Bill found in the intestine of a mosquitofish that he had just necropsied. I honestly could not believe what I saw! Subsequently, Bill identified the cestode as *Bothriocephalus acheilognathi*, known in common terms as the "Asian tapeworm." So, Bill had a model system for his dissertation research, despite my advice that there was "…nothing left in Belews Lake."

With a little literature searching, he learned that the cestode had been first described in Japan by Satyu Yamaguti in 1925 and was apparently native to China and eastern Siberia. He also discovered that the species is highly invasive and had since been found throughout Europe and North America down into Mexico. Subsequently, he was to learn that there are several reasons for its ability to colonize new localities so easily. First, it had been described from at least five different species of cyclopoid copepod first intermediate hosts, indicating very little host specificity at that level. Second, although found as adults primarily in cyprinids, it had also been reported from at least 40 different species of piscine hosts, including several noncyprinids. Third, the carp (or koi) breeding industry is very large, and the shipping of these ornamental fish throughout the world from Asia is quite lucrative. The colonization of these carp and Asian tapeworm apparently goes hand in hand with the lack of host specificity and the breeding of koi and baitfish like red shiners and fathead minnows.

The cestode is also known to cause mortality, especially in fish being raised in hatchery settings. On ingestion of copepods infected with procercoids, plerocercoids develop in the lumen of the intestine. If water temperatures are below approximately 25°C, the plerocercoid larvae stop development. When water temperature rises above approximately 25°C, these larvae transform into adult parasites. A problem arises when there are too many plerocercoids developing at the same time and blockage of the gut occurs.

The histopathology associated with *B. acheilognathi* in the gut of mosquitofish is normally minor. Occasional focal point necrosis is induced, similar to that produced by the parasite in fathead minnows as described by Scott and Grizzle (1979). Hoffman (1976, 1980) reported that large numbers of the Asian tapeworm in golden shiners and fathead minnows were the cause of intestinal obstruction and even gut perforation could cause host mortality in Arkansas and Mississippi bait farms. Evidence generated by Granath (1982) indicated *B. acheilognathi*-induced mortality occurred in Belews Lake mosquitofish, probably associated with the same phenomena described by Hoffman (1976, 1980) in his work.

The story of *B. acheilognathi* in Belews Lake probably began in 1977 when Degan and Harrell (1982) first reported fathead minnows during cove rotenone surveys routinely employed by Duke Energy to census the fish community. Mike Riggs (1986) speculated that fathead minnows reared in Arkansas fish farms and sold locally as bait were responsible for the transfer of *B. acheilognathi* into the lake, most likely an accurate conjecture. By the end of the 1970s, the main body of the reservoir was without piscivorous fish. Absent these predators, fathead minnows and mosquitofish greatly expanded their foraging ranges. For example, mosquitofish were actually captured up to 300 m from shore at the surface of water that was 40 m deep (Granath, 1982). Bill also reported that mosquitofish fed on plankton captured 300 m from shore became infected with *B. acheilognathi*.

By 1980, mosquitofish and fathead minnows were, by far, the dominant species of fish in Belews Lake. Although *Gambusia affinis* was a codominant species, fathead minnows were the primary definitive host for the cestode. Intriguingly, recruitment of *B. acheilognathi* by mosquitofish and fathead minnows displayed similar patterns in the polluted parts of the lake. Even so, parallel parasite recruitment in this case does not imply that these two host species contributed equally to tapeworm egg production in the lake. Indeed, Riggs (1986) observed just eight gravid worms from among 2247 mosquitofish necropsied over a 2-year period. Interestingly, another planktivorous fish, the red shiner, was introduced into the lake in 1980. Within a relatively short period of time, proliferation of young-of-the-year red shiners "...forced *Gambusia* populations back into the littoral zone. As a consequence, densities of *B. acheilognathi* in mosquitofish declined" and remained less than 4.0 per host for the duration of his study (Riggs, 1986).

Bill Granath's research was primarily focused on aspects of *B. acheilognathi* population biology in *G. affinis* in Belews Lake. He observed several expected, and unexpected, characteristics. There was a striking seasonal fluctuation in the prevalence and infrapopulation intensity of *B. acheilognathi*. The highest parasite intensities and prevalence were during the winter months, and their intensities, but not prevalence, then declined throughout the summer. He attributed these changes to several factors. Increasing water temperature in the spring stimulates growth and sexual maturation of the enteric *B. acheilognathi*. As they grow in size, competition for limited space occurs, and some of the parasites are lost, which accounts for high summer prevalence and the reduction in parasite intensities. These observations were also confirmed using carefully controlled laboratory experiments, which also revealed a significant affect of changing temperature on growth and maturation. Thus, no maturation occurred below 25°C, while at 30°C, nearly all of the worms had undergone segmentation. Moreover, maintenance at 25°C in the laboratory had no affect on parasite intensity, but when temperature was elevated to 30°C, the parasite intensity precipitously declined.

Another important observation made by Granath (1982) was that naturally infected mosquitofish died much sooner than uninfected *G. affinis* under similar laboratory conditions. He also established that parasite density and host size affected mosquitofish mortality. Thus, when *G. affinis* that harbored nonsegmented worms were exposed to temperatures between 25 and 35°C, small, heavily infected fish died sooner than lightly infected and larger individuals. In other words, Bill had discovered a clear case of parasite-induced host mortality and, therefore, of population regulation by the enteric *B. acheilognathi* adults in Belews Lake.

The fully operational coal-fired power plant added a considerable amount of heated water to the lake, elevating water temperatures near the plant by about 5–7°C. So, Bill set up collecting sites in both the ambient and "hot" locations of the lake. When water temperatures were high at both sites, overall parasite intensities were low, and worms were segmented. In addition, he also discovered that *G. affinis* reproduced throughout the year at the hot water site but stopped during winter at the ambient temperature site. There was a decline in parasite intensity in the smallest size class of fish at the hot site during winter, which he attributed to a dilution effect created by the absence of continuous parasite recruitment, but continued reproduction by the mosquitofish.

Bill concluded that a number of factors were involved in affecting the population dynamics of both his mosquitofish and the tapeworm in Belews Lake. Temperature, in conjunction with host feeding behavior, controlled the timing of parasite recruitment and maturation. In addition to intraspecific competition that limited parasite intensity in mosquitofish, there was also the parasite-induced host mortality.

Mike Riggs was an undergraduate student at the University of New Hampshire, obtaining his bachelor's degree in 1976. Whereas Mike's dissertation research with me began in 1979, it was interrupted when he was offered a US Public Health Service training grant award in biostatistics from the School of Public Health at UNC–Chapel Hill. He returned to Wake Forest after his degree work was completed in Chapel Hill and concluded his PhD research, graduating in 1986. Sometimes, these sorts of interruptions last a lifetime, but Mike, fortunately, persisted. Needless to say, I was very proud of his achievement!

Mike's initial goal was to use *B. acheilognathi* and *G. affinis* to do some mathematical modeling but realized that what was at first perceived to be a simple system was, in fact, far too complex to meet this objective. Part of the problem stemmed from the fact that most of the strobilate and gravid worms in the lake were associated with fathead minnows and newly introduced red shiners, not with mosquitofish. Another feature worth capitalizing on was the removal of piscine predators from the main part of the lake via the selenium pollution problem. Holmes (1979) had opined "…that any event, or series of events, which would lead to either a shifting, partitioning, or expansion of the food niches of any of the hosts would have a profound impact upon the structure of the parasite suprapopulation." He referred to this idea as the "exchange hypothesis." Since Mike felt that mathematical modeling was beyond his reach, he decided that the absence of piscine predators offered him a unique opportunity to test Holmes' so-called exchange hypothesis. Moreover, the presence of a chemocline separating the main body of the lake and the headwaters created experimental and control "patches" for comparisons (plus what turned out to be a "fluctuating interface" between the two sites). He (Riggs, 1986) was able to examine "…the multivariate effects of the physical environment, host community structure, and parasite infra-population structure on the growth rates, biomass, and fecundity of each of the component communities."

Before going any further with a discussion of Mike's work, it is necessary to briefly focus on the cyclopoid copepods and their role in the population biology of *B. acheilognathi* in Belews Lake. As noted earlier, unlike most other species of ces-todes, the Asian tapeworm exhibits very little specificity with regard to the first intermediate, or definitive, host. At least five species of copepods were found to serve as intermediate hosts in Belews Lake; three of these species are planktonic in their distribution, that is, up in the water column (*Diacyclops thomasi, Mesocyclops edax*, and *Tropocyclops prasinus*), and two are benthic, that is, associated with the lake's substrata (*Eucyclops agilis* and *Paracyclops fimbriatus poppei*). The only way copepods can become infected with *B. acheilognathi* is for them to eat free-swimming coracidia, which hatch from eggs released by gravid tapeworms. This clearly implies that the swimming behavior and seasonal dynamics of copepod populations should easily play a significant role in the population biology of the parasite. Moreover, as Mike was to discover, the differential feeding behaviors of mosquitofish, fathead minnows, and red shiners could also be an important factor in regulating the population biology of the cestode.

Mike inserted an interesting sentence in a section of his dissertation dealing with copepod biology and parasite population biology (Riggs, 1986). He stated, "Perhaps one of the most unexpected and puzzling aspects of this study concerns the seasonal periodicity of the epizootics." (By epizootics, he was referring to the seasonal peaks in prevalence and abundance of *B. acheilognathi*.) This is where the complexity of the system finally became obvious to me, as it had for Mike.

Before construction of the reservoir, Belews Lake was Belews Creek. A dam was constructed and a lake was formed—a deep lake, more than 300 ft in depth in at least one site. The volume replacement time is very slow, about 1000 days. According to Weiss and Anderson (1978), the benthos changed radically over the first 7 years of plant operations. During this change, chironomid insects were replaced by tubificid oligochaetes. In addition, a "dominant cyclopoid" community developed (Riggs, 1986). The biomass of fish stocks in the lake fell from 140 kg/ha in 1972 to 10 kg/ha in 1977 (Harrell et al., 1978). At about this time, the selenium problem in the main body of the lake was to create a depauperate fish community of three species. When piscivorous fish were eliminated, the introduction of planktivorous red shiners and fathead minnows hugely impacted the cyclopoid community structure. They also affected the feeding behavior of mosquitofish by forcing the latter into an already spatially limited littoral zone of the lake.

As any aquatic biologist would say, no two lakes are identical. Each body of water is characterized by a number of unique features. In Belews Lake, for example, there is virtually no littoral zone. As I noted earlier in my warning to Bill Granath, if you go out 15 ft from the shoreline, you will also be in 15 ft of water in many places. Moreover, when the lake was created, Duke Energy removed most of the vegetation first, that is, they eliminated all of the standing trees, except in one small cove. As a result, photosynthetic productivity in the lake was never very great because nutrient levels were so low. Moreover, farming operations and fertilizer input was relatively small in the Belews Creek watershed. On a visit to the site, George Lauff, who was Director of the W.K. Kellogg Biological Station at the time and a very good limnologist, remarked that the main body of the reservoir at least superficially resembled an oligotrophic lake, which is very unusual in the southern part of the United States—most of our impoundments are very turbid, with lots of particulates.

From the standpoint of planktonic copepod biology, *M. edax* is present only in the summer, *T. prasinus* in all seasons except winter, and *D. thomasi* during all seasons except summer (Riggs, 1986). Mature *D. thomasi* are primarily limnetic in their distribution, while their young copepodites are mostly distributed in the littoral zone of the lake. *Mesocyclops edax* in contrast are restricted to the littoral zone, and *T. prasinus* occurs in both littoral and limnetic parts of the lake. Since *B. acheilognathi* employs all three of these species as intermediate hosts, it is not difficult to see how the feeding habits and changing spatial distribution patterns of mosquitofish, fathead minnows, and red shiners affected the population biology of the Asian tapeworm over time.

Prevalence and abundance of gravid tapeworms were highest in the spring (late May and early June) when *D. thomasi* and *T. prasinus* were present in the littoral zones of the main body of the lake. In the latter part of June, prevalence and parasite abundance quickly dropped as gravid *B. acheilognathi* disappeared. Acquisition of young-of-the-year fish contributed to a decline in the parasite's prevalence. In the autumn, there was a steady increase of infection, which coincided with the emergence of infected *D. thomasi* copepodites that had spent the summer in diapause. Meanwhile, at unpolluted sites, there was no autumnal epizootic increase of *B. acheilognathi*. Why? Because piscivorous fish at these sites kept mosquitofish and fathead minnow populations relatively small and confined their distribution to the littoral zones. In polluted areas of the lake and with the disappearance of larger piscivorous predators, planktivore populations soon extended into the limnetic areas of the lake where they were able to feed on infected *D. thomasi*. Based on these observations, Riggs (1986) concluded, "Therefore, epizootic transmission could only occur in the spring and autumn in limnetic populations of fishes, but only in the spring in littoral populations."

Another question asked by Riggs (1986) was related to the effect of selenium and other factors on tapeworm fecundity in Belews Lake. He knew from observations on the fish community structure in the lake before, and during, the pollution episode that selenium in high concentrations radically affected a variety of fish species in the lake, that is, reducing the number of fish species from 24 to 3 or 4. During the course of this phase of investigation, Mike reported (Riggs, 1986) that only eight of greater than 2000 *G. affinis* necropsied possessed gravid *B. acheilognathi* and that all but one of the gravid worms were in mosquitofish seined from unpolluted sites. He attributed this observation to the "...limited confines in the gut of mosquitofish" (Riggs, 1986). In other words, space for growth and development was restricted in *G. affinis*. Indeed, he noted that mosquitofish possessing a tapeworm with 40–50 proglottids was typically "doomed" to die from the infection. However, he and Dennis Lemly also observed that nongravid, somatic tissues of tapeworms in mosquitofish accumulated much higher selenium concentrations than fathead minnows or red shiners and suggested that the selenium under these conditions could have prevented tapeworm maturation. Based on data generated in their study, Mike and Dennis (Riggs et al., 1987) asserted that selenium was having a negative effect on tapeworm fecundity in both red shiners and fathead minnows, but that "...host species is clearly the most significant influence on fecundity of *B. acheilognathi*," depending on the amount and proportion of mature strobilar tissue and rates of production and viability of egg. They concluded that "A confining environment stunts the growth of worms and reduces their capacity to mature and produce large numbers of viable eggs."

The last of the Canadians to "descend" on my lab was Dave Marcogliese. He came to Wake Forest in 1983 and received his PhD degree in 1988. His research focused on aspects of copepod community dynamics in Belews Lake prior to selenium pollution of the reservoir and during the period of greatest impact by selenium. When he arrived, we had several discussions about possible dissertation problems. He chose the copepod route because he felt that it would provide him with the best background for a career in aquatic biology and parasitology. Based on what I know to have transpired in his career since, he made the right choice.

In the early 1970s, while attending a symposium being held at the SREL, I had the opportunity of speaking with several biologists associated with the environmental division of Duke Energy. During our discussion, I mentioned that I had done some fishing out at Belews and had discovered that the largemouth bass in the reservoir were infected with *P. ambloplitis*, the bass tapeworm. Somehow, I convinced them that some research on this parasite would be a good thing for Duke Energy to invest in because of its potential to cause problems in the bass population of the lake. They provided a small grant of $15,000 or so, and as mentioned earlier, I invited one of my Biology Department colleagues, Ron Dimock, to collaborate. He agreed and we immediately began collecting plankton in the lake and saving the copepods for research that we would do later. Neither of us suspected at the time that a Canadian named Dave would be interested in a fairly large collection from preselenium Belews Lake, but it worked out that way. So, the plankton collected by Ron and me became Dave's preselenium, or control, samples, and he was off and running on his dissertation research.

While some of Dave's work was focused on *B. acheilognathi* in Belews Lake, most of it was confined to a study of plankton community biology in response to the absence of piscivorous fish due to selenium poisoning (Marcogliese, 1988). We knew that 24 species of fish were present in the lake prior to the selenium problem and that four species (three residual species and one colonizer) were all that were present by 1976–1977. Prior to that time, the presence of piscivorous fish species, primarily largemouth bass, kept smaller planktivorous species restricted to littoral zones of the lake. As a consequence, physically larger plankton species were dominant in the limnetic waters. However, with the elimination of piscivorous fish, planktivores (mainly mosquitofish at first), followed by fathead minnows and then red shiners, moved into limnetic parts of the lake. Interestingly, diversity in the plankton community did not change, but the size of the dominant species did.

The results of Dave's work on this aspect of his dissertation offered support for the Menge–Sutherland hypothesis, which suggested "…that predation can regulate species composition at lower trophic levels" (Marcogliese, 1988). The removal of piscivores allowed for increased planktivory and forced a change toward increased abundance of smaller planktonic copepod species. According to Dave, "These second and third order effects are precisely what has been predicted by the models of cascading trophic interactions in aquatic systems (Carpenter et al., 1985; Kerfoot, 1987)." He (Marcogliese, 1988) concluded by stating that in Belews Lake prior to selenium pollution, "…the top carnivore was represented by an assemblage of piscivorous fish. Their presence permitted the maintenance of substantial populations of large zooplankters within the zooplankton community."

Belews Lake is configured in such a way that heated effluent from the steam station is initially confined to a "cooling arm" that is, in turn, connected by a canal to the main body of the reservoir. By the time the water is transferred from the cooling arm back into the main body, its temperature is almost the same as in the main body of the reservoir. Water temperatures at the source of the heated effluent were, on average, approximately 5–7°C higher than at ambient sites. An analysis of copepod community structure at the heated and ambient temperature sites showed that diversity patterns were not different, a clear indication that temperature was not affecting direct changes in the planktonic community structure.

Many organisms in thermally stratified freshwater and marine ecosystems exhibit vertical migrations on a diurnal basis. Explanations for this behavior are varied, with some attributing it to a means for escaping visual predators, while others view it as being a metabolic or demographic benefit by reducing metabolic demands at elevated temperatures. In Belews Lake, the thermocline at the beginning of the 1986 summer was between 7 and 11 m (Marcogliese, 1988). Water intake valves for the steam generation at the plant are located at 9 m. In order to increase cooling efficiency, Duke Energy installed a pair of air injectors at 17 m. The idea was to mix cooler hypolimnetic waters with water from the epilimnion that would in turn cool the warmer water and increase the plant's efficiency. In accomplishing this goal, the hypolimnion dropped below 17 m. Dave's idea was:

If migration occurs solely to avoid visually oriented predators, zooplankton must move into the thermocline or below it. However, if migration occurs solely to avoid visually oriented predators, zooplankton need only to migrate out of the euphotic zone in daylight. If metabolic or demographic gains are adaptive, animals will cue on temperature and descend into the thermocline or below it. (Marcogliese, 1988, p. 104)

He observed the diurnal patterns of movement for cyclopoids prior to restratification in Belews and found that two of the species migrated vertically. *Mesocyclops edax* was a strong migratory copepod before and after restratification, following the cooler water down to 15–20 m. A second species, *T. prasinus*, was present in the top 10 m during the day but moved upward at night. *Diacyclops thomasi* did not exhibit any migratory activity.

His results suggested that

This study provides the first empirical field data supporting the metabolic or demographic models of vertical migration (McLaren, 1963, 1974). Though predator avoidance can be of primary importance in influencing diurnal vertical migration in unstratified systems (Zaret and Paine, 1973), it is now clear that migrations in more structurally complex systems cannot be satisfactorily explained by comparable mechanisms. (Marcogliese, 1988, p. 110)

Belews Lake has now returned to its original state. We cannot say whether the original community of 24 species has returned, but we suspect that it has. We know that there were refugia for these species in the headwaters of both the ambient and cooling parts of the reservoir. A few years ago, and several years after the pollution problem was resolved, anglers fishing in the main body were still warned not to consume any of the fish caught because selenium concentrations in their flesh was too high. However, selenium release into the system was stopped in the late 1980s when Duke Energy began burning coal from other sources. It is our understanding that selenium in the water was being reduced gradually by natural precipitation and incorporation into the bottom substrata. Gradually, over time, these substrata have been covered over via natural processes, and the lake water is now free of selenium.

We have not examined the fish community structure in the reservoir since Dave Marcogliese's work in the 1980s. However, Browne and Lutz (2010) published the results of a very thorough study in which they carefully scrutinized the zooplankton and piscine communities. They observed that "As the piscivore population in Belews Lake returned to its pre-impact density, macrozooplankton recovered to baseline levels of 1974–1975 in terms of density and species diversity, as measured by both raw counts and Shannon-Wiener indices."

Belews Lake and Charlie's Pond are about 20 miles from Wake Forest but only approximately 20 m from each other. The research opportunities these two bodies of water provided my students from 1978 to the present time were truly remarkable. The effort in both of them is now over because I doubt that I will be taking on any more students. However, I greatly hope that at a point in the future, some young parasitologist with an interest in ecology will say to themselves, "Belews Lake and/ or Charlie's Pond need to be looked at again, just to see how the systems have changed since Esch's crew worked out there." Long-term studies of this sort are rare, but they present a really good chance to do some exciting research. My students have created a very large opportunity for someone.

References

Browne, R.A., and D. Lutz. 2010. Lake ecosystem effects associated with top-predator removal due to selenium toxicity. *Hydrobiologia* **65**: 137–148.

Carpenter, S.R., J.F. Kitchell, and J.R. Hodgson. 1985. Cascading trophic interactions and lake productivity. *BioScience* **35**: 634–639.

Degan, D., and D. Harrell. 1982. 1978–1981 cove rotenone sampling of Belews Lake. Duke Power Co. Research Report N. EL/82-19. Duke Power Co., Charlotte, North Carolina.

Goater, T.M., C.P. Goater, and G.W. Esch. 2013. Parasitism: The diversity and ecology of animal parasites, 2nd edition. Cambridge University Press, Cambridge, UK.

Granath, Jr., W.O. 1982. Studies on the biology of Bothriocephalus acheilognathi (Cestoda: Pseudophyllidea) in mosquitofish, Gambusia affinis. Ph.D. dissertation, Wake Forest University, Winston-Salem, North Carolina, 179 p.

Harrell, R.D., R.L. Fuller, and T.J. Edwards. 1978. An investigation of the fish community in Belews Lake, North Carolina. Duke Power Company tech. Rep. Ser. No. 78-07. Duke Power Company, Charlotte, North Carolina.

Hoffman, G.L. 1976. The Asian tapeworm, Bothriocephalus gowkonensis, in the United States and research needs in fish parasitology. Proceedings of the 1976 Fish Farming Conference, Annual Convention of Catfish Farmers of Texas, Texas Agricultural and Mechanical University, College Station Texas, p. 84–90.

Hoffman, G.L. 1980. Asian tapeworm, *Bothriocephalus acheilognathi* Yamaguti 1934, in North America. *Fisch und Umwelt* 8: 69–75.

Holmes, J.C. 1979. Parasite populations and host community structure. *In* Host-parasite interfaces, B.B. Nickol (ed.). Academic Press, New York, p. 27–46.

Kerfoot, W.C. 1987. Cascading effects and indirect pathways. *In* Predation: Direct and indirect impacts on aquatic communities, W.C. Kerfoot, and A. Sih (eds.). University of New England Press, Hanover, NH, p. 57–70.

Marcogliese, D.J. 1988. Studies on the plankton community of Belews Lake, North Carolina, with a discussion of the role of cyclopoid copepods in transmission of the fish tapeworm, Bothriocephalus acheilognathi. Ph.D. dissertation, Wake Forest University, Winston-Salem, North Carolina, 214 p.

McLaren, I.A. 1963. Effects of temperature on growth of zooplankton and adaptive value of vertical migration. *Journal of the Fisheries Research Board of Canada* 20: 685–727.

McLaren, I.A. 1974. Demographic strategy of vertical migration by a marine copepod. *American Naturalist* 108: 91–102.

Riggs, M.R. 1986. Community dynamics of the Asian tapeworm, Bothriocephalus acheilognathi, in a North Carolina cooling reservoir. Ph.D. dissertation, Wake Forest University, Winston-Salem, North Carolina, 340 p.

Riggs, M.R., A.D. Lemly, and G.W. Esch. 1987. The growth, biomass, and fecundity of *Bothriocephalus acheilognathi* in a North Carolina cooling reservoir. *Journal of Parasitology* 73: 893–900.

Scott, A.L., and J.M. Grizzle. 1979. Pathology of cyprinid fishes caused by *Bothriocephalus acheilognathi* Yeh, 1955 (Cestoda: Pseudophyllidea). *Journal of Fish Diseases* 2: 69–73.

Weiss, C.M., and T.P. Anderson. 1978. Belews Lake. A summary of seven years of study to assess the environmental effects of a coal-fired power plant on a cooling pond. ESE No. 475. Department of Environmental Science and Engineering, School of Public Health, University of North Carolina, Chapel Hill, North Carolina.

Zaret, T.M., and R.T. Paine. 1973. Species introduction into a tropical lake. *Science* 182: 449–455.

9 The Strigeids

Where there is much desire to learn, there of necessity will be much arguing, much writing, and many opinions, for opinion in good men is but knowledge in the making.

Areopagitica, John Milton (1608–1674)

Over the years, students in parasitology labs around the world have happened upon a number of strigeid trematodes while working with snails and fish from the field. Most of these encounters have been accidental, and their reporting has been largely anecdotal. However, three of my graduate students successfully focused their Master's or PhD research on strigeid trematodes. For example, John Aho and Joe Camp worked on these flukes down at the Savannah River Plant (SRP) in Par Pond, South Carolina, the greater than 1100 ha cooling reservoir described in Chapter 3. Two of these trematodes were *Tylodelphys (Diplostomulum) scheuringi* (*Ts*) and *Ornithodiplostomum ptychocheilus* (*Op*). The life cycles of the two species are similar to those of most other strigeids in that snails serve as first intermediate hosts, that is, *Helisoma (Hel.) anceps* for the former species and *Physa gyrina* for the latter. The primary second intermediate hosts for both trematode species in Par Pond are mosquitofish, *Gambusia affinis*, with several species of fish-eating birds serving as definitive hosts. The sites of infection for the two flukes in their second intermediate hosts are, however, very different. In the case of *Ts*, unencysted metacercariae occur in the body cavity. Metacercariae of *Op* are encysted and can be found in the cranial cavity, ocular orbits, and eyes. The two species of snail hosts and mosquitofish are sympatric in both the heated and ambient temperature sites of the reservoir.

Dennis Lemly focused on a third strigeid, *Uvulifer ambloplitis*, the cause of black spot or black grub (Figure 9.1). The first intermediate host for *U. ambloplitis* in our "neck of the woods" is *Hel. anceps*. Furcocercous cercariae released from the snail

Ecological Parasitology: Reflections on 50 Years of Research in Aquatic Ecosystems,
First Edition. Gerald W. Esch.
© 2016 John Wiley & Sons, Ltd. Published 2016 by John Wiley & Sons, Ltd.

Figure 9.1 *Lepomis macrochirus* with black spot, each of which is a metacercaria of *Uvulifer ambloplitis*. Photograph by Dennis Lemly.

penetrate the surface of a number of piscine hosts, but centrarchids (sunfish) are apparently the favored second intermediate hosts. On penetration, the tails are shed, and the parasite produces a cyst wall and becomes sequestered. The cyst wall and/or secretions from the metacercaria stimulate a strong reaction on the part of the host. Melanocytes and fibrocytes are drawn to the metacercaria, and a very thick and darkly pigmented and fibrotic wall develops around the parasite, hence the name black spot or black grub. The definitive hosts are kingfishers (*Megaceryle* spp.).

I met John Aho in 1973 when he was a sophomore student here at Wake Forest; he was taking one of my lower division courses. John also worked for me as a lab assistant while he was an undergraduate and then briefly for Whit Gibbons as a research technician down at the SRP. I was delighted that he decided to undertake study as a Master's student with me here at Wake Forest, even though all of his fieldwork would be done at the SRP.

John was interested in the strigeid/mosquitofish system because of his experience at the SRP over a period of several years. As described elsewhere, I had been working down there since the early 1970s and had involved my students with field studies dealing with aspects of thermal ecology and its effects on parasite population biology in both largemouth bass and in turtles. John had become interested in the strigeid system while working in Par Pond and several other aquatic field sites in the vicinity of the big reservoir. Thermal biology was a very "hot" research focus in ecology at the time, primarily because of the large number of nuclear and coal-fired power stations that was being constructed in those years. Biologists were not exactly certain what impact elevated water temperature would have on the biota in aquatic habitats. Joe Camp became interested in the field site and strigeids because of the work John had accomplished. Joe continued John's line of research in Par Pond. Their research thus greatly complemented each other and worked out quite well.

The only difficulty was that Joe was about a 150 miles away from the SRP, SREL, and Par Pond. The fieldwork he was obliged to do required regular visits to the site, about a 3-hour drive between the site in South Carolina and Winston-Salem, the home for Wake Forest University. Because of the distance between here and SREL, my personal admiration for Joe and his tenacity grew by "leaps and bounds" as time passed.

As I have indicated, work in thermal ecology was extensive during the late 1960s and into the 1970s. It covered a wide range aquatic habitats and species. The SRP site presented exceptional opportunities because the temperatures were as high as 95°C when water was discharged from the reactors into stream and reservoir settings. It should be noted that most states, including South Carolina, have specific laws that limit elevated temperatures in reservoirs, which are exposed to thermal effluents. These laws, however, do not pertain to federal properties such as the SRP. During the study periods for Aho (1979) and Camp (1980) in the heated areas of Par Pond, the maximum water temperature reached was 41°C. The maximum ambient temperature in unaltered areas of Par Pond was 33°C. As expected, water temperatures varied seasonally, but the widest ranges were seen in the heated areas because of periodic fluctuations in reactor operations at the plant.

One might guess a priori that elevated temperatures would increase cercariae production by snails and perhaps stretch out periods of parasite recruitment by mosquitofish. This is what did occur at the heated end of the reservoir and for both species of fluke when compared with parasite shedding activity in the ambient area of Par Pond. However, there was also measurable inconsistency in the patterns of parasite recruitment by *G. affinis* over time, especially at the heated end of the reservoir. At the outset of the studies, the causes for these variations had not been determined. After a few months into Joe's work, however, he began using a live-box procedure for gathering a more accurate estimate of timing for parasite recruitment by mosquitofish. This method involved the positioning of wire "cages" (sometimes also referred to as tethers) containing mosquitofish at both study sites in the reservoir for varying lengths of time. Uninfected fish were obtained from abandoned farm ponds on the SRP and placed in the "boxes" to obtain an accurate picture of parasite recruitment by the piscine second intermediate hosts. One notable feature identified by this technique was the occurrence of breaks in recruitment during periods of the summer when water temperature at heated sites increased to exceptional heights. It was also noted that during these periods, infected snails were very difficult to find.

During the last 30 months of Joe's study, the reactor operations were highly irregular, which meant that water temperatures varied greatly. There were times when they would climb above ambient temperatures by as much as 9°C. Under these conditions, the erratic temperature suggested the possibility of inducing thermal shock, or stress, which could be conducive to reducing the efficiency of parasite transmission through snails to the mosquitofish.

There was also some evidence generated that suggested the growth and life span of mosquitofish were affected by elevated temperatures, both of which would significantly impact parasite intensities and prevalence. For example, in November 1976, nearly 20% of the mosquitofish at the ambient site were greater than 40 mm in length, while none was greater than 40 mm at the heated site. In April 1977, approximately 40% were greater than 40 mm at the ambient site, but none was greater than 40 mm at the heated site. These observations suggested that *G. affinis* matured more rapidly in heated locations and then senesced and died at smaller sizes. Modifications in these temperature diversions caused much disruption in parasite recruitment patterns during these years.

While I am obviously biased, I feel that one of the best series of studies dealing with parasite frequency distributions and their application to understanding parasite population biology was conducted by Dennis Lemly. His work (Lemly, 1983; Lemly and Esch, 1984a, 1984b, 1984c) was focused on the population biology of the strigeid, *U. ambloplitis*, and the causative agent for black spot disease in centrarchid fishes.

While pursuing an MAEd at Wake Forest, Dennis and his biology faculty advisor (Dr John Dimmick) conducted a considerable amount of fieldwork searching for bluegill (*Lepomis macrochirus*) ponds in the area around Winston-Salem. Not surprisingly, they also identified a number of impoundments where strigeid metacercariae were found to infect fish. However, in one of them, Reed's Pond, located in Davidson County, just south of Winston-Salem, Dennis observed that the bluegill sunfish were heavily infected with black spot metacercariae. These metacercariae, or "black grubs" (Figure 9.1), were first described by Hughes in 1927, and the life cycle of *U. ambloplitis*, the causative agent for black spot, was resolved by Krull (1934).

When Dennis began research on bluegill sunfish, he initially noted that very little new information regarding black spot disease had emerged since the parasite was first described and the life cycle was resolved. However, during his master's degree work, he discovered that bluegills in Reed's Pond with more than 50 metacercariae were very common in the fall months but that by the following spring, the heavily infected fish were gone. This suggested that fish with greater than 50 cysts could be dying over the winter or that the parasites were dying and the cysts were being resorbed. In view of the fish mortality, that is, 10–20% among young-of-the-year fish, plus the absence of ecological information regarding the population biology of *U. ambloplitis*, Dennis decided to concentrate on the black spot parasite as the central part of his dissertation research. He narrowed his objectives, finally deciding to follow the seasonal population biology of the parasite in both snails and bluegill sunfish, examine the ability of black spot to survive over winter in the sunfish, test the hypothesis regarding the mortality of heavily infected fish, and determine if mortality was being caused in bluegills over winter and, if so, what was (were) the cause(s).

The field data generated in Lemly's (1983) study covered most of 1979, all of 1980 and 1981, and most of 1982. Also included were results from a wide range of laboratory experiments designed to answer a variety of specific questions raised as he collected data from fieldwork over the nearly 4-year period of research.

Reed's Pond represents a classic example of what is referred to by the "locals" as a bass–bluegill impoundment. Farmers make use of these ponds for watering livestock year-round. But they also stock them with largemouth bass and bluegill and use them for recreational purposes as well. The "trick" in suitable stocking is to obtain the proper ratio of bluegill and bass to prevent stunting of the latter species. In the case of Reed's Pond, the ratio was apparently a good one because a stable number of healthy bass and bluegill were present throughout the study period. Interestingly, representatives of both species sustained excellent body condition throughout the study even though fishing pressure had been nil for the previous 20 years (Lester Reed, personal communication [from Lemly, 1983]). Dennis' explanation for the consistency will become apparent as the results of his research are discussed.

His observations in the pond indicated that metacercariae recruitment by young-of-the-year bluegills began in April and continued throughout the summer before stopping in October. During this period, the variance/mean ratios of black spot metacercariae continuously grew larger but then stopped increasing with cessation of parasite recruitment in October, as expected. The increase in these ratios is an indicator of the extent of aggregation by metacercariae in the bluegill population. If parasite recruitment ceased in October, then it was surmised that the ratios should have remained constant throughout the winter months. However, this did not happen. After 3 months, the ratios were 10- to 100-fold smaller. This same sort of observation had been made for several other host–parasite systems (e.g., Crofton, 1971; Gordon and Rau, 1982). As an explanation for these findings, the latter authors, and others, hypothesized that parasite-induced host mortality had caused the declines, but testing the hypotheses by these investigators was not attempted. This is one of the features that separated Dennis' work from several other similar efforts.

Most studies of this kind attempted to also do some "curve fitting" using their dispersion data. Similar to Crofton's results (1971), the negative binomial model adequately described the metacercariae frequency distribution in bluegill sunfish. In Lemly's (1983) results, "overdispersion resulted in a highly skewed distribution of parasites among the bluegill in Reed's Pond, with 80% of the cysts present in only 23–47% of the hosts."

Lemly and Esch (1984c) suggested that the only way to account for such changes in the variance/mean ratios would be for the bluegill population to recruit uninfected fish or for there to be mortality within the bluegill population in the pond. The former explanation was immediately rejected because the pond was isolated so that colonization by uninfected fish could not occur, leaving bluegill mortality as the only plausible explanation.

He used several approaches to test his hypothesis, the results of which consistently supported his ideas regarding density-dependent, parasite-induced host mortality. He observed that any young-of-the-year fish with 50, or more, metacercariae died when water temperatures reached, or dropped below, 10°C. Dennis estimated that 10–20% of the bluegill population was eliminated in the winter months of each year as a direct consequence of high parasite intensity and cold water temperatures. It should be added at this point that feeding by bluegill sunfish almost completely stops in winter. Accordingly, in order to survive, the nonfeeding bluegill would require exploitation of stored energy reserves, primarily lipids. He also sampled largemouth bass in the summer to determine if there might be selective predation on heavily infected bluegills, but the data were negative with respect to this hypothesis.

Body condition (K) is a body length–weight metric used by fish biologists and others as an indicator of robustness or fitness (not genetic fitness). Throughout the period of parasite recruitment, there was a direct correlation between body condition, total body lipid, and parasite density in the bluegills. As K increased, body condition increased. However, as parasite density increased, both body condition and total lipid declined. Based on several studies by other investigators (Hunter and Hamilton, 1941; Erasmus, 1960), Dennis suggested that there was a very strong energy demand induced by newly acquired parasites, most likely because of inflammation and melanization processes that occurred during metacercariae recruitment and encystation. He believed that infection by *U. ambloplitis* metacercariae was a high energetic cost factor and should stimulate an increase in oxygen consumption by the fish. He confirmed this idea by observing that O_2 consumption in fish held at 25°C was significantly higher at 30 days postinfection and that it had returned to normal by 60 days postinfection.

It was mentioned earlier that he (Lemly, 1983) was attracted to Reed's Pond by the observation that young-of-the-year fish with more than 50 metacercariae were not surviving the winter. He checked fish maintained in the laboratory under ambient temperature conditions to be certain that some of the metacercariae were not dying and being resorbed during the winter months. However, consistent metacercariae numbers persisted throughout the winter months. And, as indicated previously, heavily infected bluegills were not being selectively preyed upon by largemouth bass. He also noted that kingfisher predation was mainly a summer phenomenon, coinciding with periods of nesting. He thus determined that neither predator was preferentially selecting heavily infected fish. As the study progressed, he was able to observe behavior of uninfected fish and bluegills with heavy infections, but he was unable identify any differences in the two groups of bluegill.

Depletion of stored lipid and cold temperatures (<10°C) went "hand in hand." He also found that fish greater than 70 mm in total length were unaffected by infections with more than 50 metacercariae. His explanation was that fish of the larger size moved away from the littoral zone into deeper waters and that "…consequently, their parasites were acquired when the fish were smaller and frequented shallower water" (Lemly, 1983).

Another interesting surmise by Dennis had to do with the balance in numbers of bluegill and largemouth bass in the pond. As noted previously, the bass–bluegill ratios in Reed's Pond had been in good balance for at least 20 years. In many populations, the ratio becomes uneven, with an increase in the size of the bluegill population and stunting if the bass population is "overfished." Despite the fact that Reed's Pond was not regularly fished for bass over a number of years, the two fish species seemed to be in good balance throughout the study. Dennis suggested that the annual removal of 10–20% of the bluegill population may have influenced this facet of the bass–bluegill ratio and that black spot disease actually helped to "...crop the excess in reproduction that could lead to overcrowding, stunting, and population imbalance commonly associated with the bass-bluegill combination in small lakes and ponds" (Lemly, 1983).

I must add a brief postscript here. Dennis Lemly and Mike Riggs (see Chapter 8) became close friends and even research colleagues during their time here at Wake Forest. In fact, if one returns to Chapter 8 and refers to a Riggs et al. (1987) citation having to do with growth, biomass, and fecundity of *Bothriocephalus acheilognathi* in Belews Lake, half of the "et al." in that citation refers to Dennis Lemly (I am the other half). Dennis' interest in selenium was spurred when he and his early faculty advisor sampled Belews Lake and found no bluegill sunfish in the main body of the reservoir, a most unusual finding since this species is probably the most abundant and widespread centrarchid in the southeastern United States. Dennis contacted the environmental people at Duke Energy and was told that only four species remained in the main body of the reservoir (a few carp and catfish and many mosquitofish and fathead minnows) of the 24 species that had been present just a few years earlier. The elimination of 20 species from the main body of the reservoir was due high concentrations of selenium that came from coal fly ash (see Chapter 8).

Dennis' interest in selenium toxicity was pushed to the extent that he actually took a course in analytical chemistry at Wake Forest while conducting his black spot research so that he could learn how to measure selenium levels in the bluegills. Dennis and Mike Riggs also measured selenium concentrations in the strobilae of *B. acheilognathi* from both the polluted and nonpolluted areas of the reservoir. They found that selenium concentrations in the tapeworm were exceptionally high and that the egg production by *B. acheilognathi* was significantly reduced in these worms. This finding coincides with previous observations that relatively small quantities of selenium were sufficient to cause sterility in piscine species.

When Dennis finished his PhD, he headed for a 3-year postdoctoral position at the University of Saskatchewan and the laboratory of Professor R.J.F. Smith so that he could pursue research on water quality. He had made a career decision that was to take him into the Forestry Service and subsequently the Department of Fisheries and Wildlife in the US government, where he continued his research on the ramifications of fly ash and selenium pollution. He is now recognized as a world-class expert in both areas.

The work performed by John, Joe, and Dennis was elegant and has received solid attention in the intervening years. As with the other students I have written about, their work was revealing and consequential in the area of parasite ecology. Even though Joe and John worked with completely different parasite and host taxa, their efforts were complementary because they were focused on the same sort of environmental stress. Dennis' research was highly distinctive because he actually tested some of the then new quantitative ideas proposed by Crofton (1971).

References

Aho, J.M. 1979. Thermal loading and parasitism in the mosquitofish, Gambusia affinis. M.S. thesis. Wake Forest University, Winston-Salem, North Carolina, 110 p.

Camp, Jr., J.W. 1980. Studies on the population biology of Diplostomulum scheuringi in mosquitofish, Gambusia affinis. Ph.D. dissertation. Wake Forest University, Winston-Salem, North Carolina, 111 p.

Crofton, H.D. 1971. A quantitative approach to parasitism. *Parasitology* **62**: 179–193.

Erasmus D.A. 1960. The migration of *Cercaria X* Baylis (Strigeidae) within the fish intermediate host. *Parasitology* **49**: 173–190.

Gordon, D.M., and M.E. Rau. 1982. Possible evidence for mortality induced by the parasite *Apatemon gracilis* in a population of brook sticklebacks (*Culaea inconstans*). *Parasitology* **84**: 41–47.

Hughes, R.C. 1927. Studies on the trematode family Strigeidae. IV. A new metacercaria, *Neascus ambloplitis*. *Transactions of the American Microscopical Society* **46**: 248–267.

Hunter, G.W., and J.M. Hamilton. 1941. Studies on host-parasite reactions to larval parasites. IV. The cyst of *Uvulifer ambloplitis* (Hughes). *Transactions of the American Microscopical Society* **60**: 498–507.

Krull, W.H. 1934. *Cercaria bessiae* Cort and Brooks, 1928, an injurious parasite of fish. *Copeia* **1934**: 69–73.

Lemly, A.D. 1983. Ecology of Uvulifer ambloplitis (Trematoda: Strigeidae) in a population of bluegill sunfish, Lepomis macrochirus (Centrarchidae). Ph.D. dissertation. Wake Forest University, Winston-Salem, North Carolina, 225 p.

Lemly, A.D., and G.W. Esch. 1984a. Population biology of the trematode, *Uvulifer ambloplitis* (Hughes, 1927), in the snail intermediate host, *Helisoma anceps*. *Journal of Parasitology* **70**: 461–465.

Lemly, A.D., and G.W. Esch. 1984b. Population biology of the trematode *Uvulifer ambloplitis* in juvenile bluegill sunfish, *Lepomis macrochirus*, and largemouth bass *Micropterus salmoides*. *Journal of Parasitology* **70**: 466–474.

Lemly, A.D., and G.W. Esch. 1984c. Effects of the trematode *Uvulifer ambloplitis* on juvenile bluegill sunfish, *Lepomis macrochirus*. Ecological implications. *Journal of Parasitology* **70**: 475–492.

Riggs, M.R., A.D. Lemly, and G.W. Esch. 1987. The growth, biomass, and fecundity of *Bothriocephalus acheilognathi* in a North Carolina cooling reservoir. *Journal of Parasitology* **73**: 893–900.

10 Some Small Streams and Small Ponds

Rivers perhaps are the only physical features of the world that are at their best from the air.... Rivers stretch out serenely ahead as far as the eye can reach.
North to the Orient, Anne Morrow Lindbergh (1907–2001)

Mike Barger was introduced to parasitology at the University of Nebraska–Lincoln (UNL) by John Janovy, Jr., an early graduate school companion of mine. Mike had enrolled in an introductory zoology course taught by John during the second semester of his freshman year. After scoring an 88 on his first zoology hour exam, Mike told me that he was approached by John and asked to report to his office within 30 days. Thirty days later, he complied (Mike did not say why he waited for the entire 30 days). On meeting with John, he was invited (recruited) to work in his lab as an undergraduate research student. This meant that a real good research opportunity was waiting for him. One of the many things that both John and I learned from Dr. Self at Oklahoma while we were his grad students was not to use undergrads as "gofers" in your research lab (by a gofer, I mean someone whose only job is to "go for this," or "go for that," or just wash dishes; a "gofer" does little more than plain "scut" work). Consequently, Mike published his first paper with John based on an undergraduate research project conducted at the Cedar Point Biological Station in western Nebraska. It was this wonderful "hands-on" experience with John, and subsequently Brent Nickol, that swayed Mike into pursuing an academic career.

Toward the end of his undergraduate days, a young faculty member at UNL attempted to convince Mike that he should leave Nebraska and do his graduate work elsewhere. Mike told the "ad hoc" advisor that he did not want to leave UNL, so the

Ecological Parasitology: Reflections on 50 Years of Research in Aquatic Ecosystems, First Edition. Gerald W. Esch.

same guy recommended that the least he should do was to leave John (apparently to gain some mentor diversity) and adopt Brent Nickol as his new guide/counselor, which he did (no, the faculty advisor was not Brent). So, after switching to Brent's lab, Mike did the research necessary for his MS degree and then came to Wake Forest for his PhD dissertation.

During a discussion we had soon after he arrived here, Mike suggested a couple of ideas about possible dissertation directions he should take. One of them was to expand some work that Joe Bourque and Kym Jacobson had accomplished down at the Savannah River Ecology Laboratory (SREL) with acanthocephalans in slider turtles. However, I dissuaded him from taking that route because of the extensive traveling he would need to do (I recalled all of the long trips that Joe Camp had to make when he worked on his dissertation at SREL and felt Mike could and should find something closer to Winston-Salem). Mike also indicated an interest in following up on Tim Goater's Master's research involving salamander-parasite ecology in the streams of the Great Smoky Mountains of southwestern North Carolina. In a recent conversation, Mike reminded me that I had discouraged him from pursuing that too, again because of the long travel that would be required.

Just before Mike arrived at Wake Forest, my wife, Ann, and I traveled to England and Ireland. One of the people I wanted to see was Celia Holland at Trinity College in Dublin. I recall talking with her about the dissertation research she had accomplished dealing with soil-transmitted nematodes in humans in Africa. She described how she had used one of the approaches proposed by Anderson and May for treating of geohelminths in an African village. It was at this point I wondered if the same methodology could be employed with *Halipegus (=Hal.) occidualis* in Charlie's Pond, North Carolina. In some ways, our system was even better than the one used for humans since we could manipulate the infrapopulation numbers of *Hal. occidualis* in frogs with much better precision than the enteric geohelminths in humans. If one wanted 10 adult *Hal. occidualis* in a green frog and there were 13 present, all you had to do was open the mouth of the frog, lift the tongue, and remove three parasites; if there were seven present, then open the mouth, lift the tongue, and add three. It was easy. Obviously, with human geohelminths, this procedure was not possible. I remember e-mailing Derek Zelmer (see Chapter 6) from Celia's office immediately after talking with her and telling him not to start his fieldwork in the streams of western North Carolina yet that I thought there was a better dissertation project using the *Hal. occidualis* system in Charlie's Pond.

The research we had first projected for Derek would have taken him into the mountain streams not far from Winston-Salem where he was going to examine aspects of the population and community ecology of parasitic helminths in stream fishes. However, because of my talk with Celia, we came up with something different for Derek, and when Mike arrived, the field study in the mountains went to Mike. I think, in the end, both Mike and Derek were satisfied with their problems and their outcomes.

On the east coast of the United States, there is a stream known as the Great Pee Dee River, which empties into the Atlantic Ocean at Georgetown, South Carolina, about halfway between Myrtle Beach to the north and Charleston to the south. Recognized as one of the longest drainage basins on the east coast of the United States, the Pee Dee actually begins as the Yadkin River in the Appalachian Mountains near the tourist community of Blowing Rock, North Carolina. The Yadkin changes to the Pee Dee in southern North Carolina where the former is joined by the Rocky River. When Derek decided to go with the *Hal. occidualis* system in Charlie's Pond, he volunteered to take Mike on a camping trip into the upper Yadkin River region to scout several of the mountain streams as possible research sites. They seined for fish and collected any snails they could find in the same creeks. The fish and snails were returned to Wake Forest where they were necropsied and checked for parasites.

At that point in time, Mike had no a priori ideas regarding how to approach his problem, but it was to "hit" him very quickly. He eventually focused his research efforts on the two most abundant fish species, that is, rosyside dace (*Clinostomus funduloides*) and redlip shiners (*Notropis chiliticus*). The parasite component community in the former included 11 species, with 6 species in the latter host.

The first paper generated by Mike after arriving at Wake Forest (Barger and Esch, 1999), however, was the description of a new opecoelid species, *Allopodocotyle chiliticorum*, isolated from redlip shiners, but it had nothing to do directly with the stream ecology on which he was to focus for his dissertation. His second publication also followed quickly. It too was totally unique and quite unexpected (Barger and Esch, 2000). *Plagioporus sinitsini* is another opecoelid fluke that occurs as an adult in rosyside dace in one of Mike's mountain streams (Basin Creek). It can employ a three-host life cycle, typical for most trematodes. Maturation of eggs shed from dace requires about 14 days to complete, at which time they hatch and miracidia infect the prosobranch snail, *Elimia symmetrica*. Subsequently, cotylocercous cercariae are shed from the snail. Over the course of his work on *P. sinitsini*, these cercariae were observed to penetrate various mayfly and stonefly naiads, where metacercariae then developed.

However, on searching the literature, Mike discovered that *P. sinitsini* could also exhibit a truncated life cycle, that is, daughter sporocysts are voided from the alimentary canal of *E. symmetrica* snails found in Michigan (Dobrovolny, 1939). The shed sporocysts were observed to possess both cotylocercous cercariae and metacercariae; consumption of the sporocysts by the definitive hosts completed the life cycle.

On transporting *E. symmetrica* snails to the lab, Mike would routinely isolate them in small plastic jars containing aged tap water to check for cercariae that might be released. While following this protocol, Mike noted the presence of what he at first thought was fecal material in the plastic snail jars. However, he quickly observed that these masses did not resemble typical fecal droppings, so he examined them more carefully using a compound microscope. He was surprised to find not only shed sporocysts but also cotylocercous cercariae and adult trematodes inside the sporocysts. Moreover, the adults were shedding eggs while still inside sporocysts. A very thin and translucent cyst wall enclosed some of the adults inside the sporocysts, while others were entirely free. He collected adults from both dace and adults from the shed sporocysts and found them to have identical morphology and metric dimensions. Eggs from adult flukes isolated from dace and sporocysts hatched after a 14-day maturation period, releasing active miracidia. He also noted that some snails initially shed cercariae, then stopped and began shedding sporocysts with cercariae. Other snails that were initially shedding sporocysts stopped and began releasing cercariae. In other words, he found that the life cycle of *P. sinitsini* was capable of using three hosts but that it could also be abbreviated to the point of a one- or two-host life cycle as well.

A phenomenon known as paedogenesis involves the retention of juvenile characteristics by adults. It "...results in retarded somatic development relative to development of the germ line (Raf, 1996) and has been hypothesized to be a common cause of life cycle truncation among parasites (Brooks and McLennan, 1993)" (Barger and Esch, 2000). Initially, the situation for *P. sinitsini* was thought to be a form of paedogenesis known as progenesis, that is, early onset of sexual maturity. A definition of progenesis provided by Gould (1977) (as repeated in Lefebvre and Poulin (2005)) is, "Any heterochronic development in which first reproduction occurs at an earlier age, that is, sexual reproduction in an organism still in a morphologically juvenile stage." The latter authors noted the definition by Gould (1977) does not match the situation that is typically seen in many trematodes, including *P. sinitsini*, that is, "...the production of viable eggs in individuals inhabiting an organism that would normally be considered an intermediate host" (Lefebvre and Poulin, 2005).

Based on a careful search of the existing literature at the time of his publication (Barger and Esch, 2000), Mike concluded that the *P. sinitsini* system in Basin Creek exhibited some rather unique qualities. Lefebvre and Poulin (2005) had previously provided a thorough and extensive overview of progenesis in a variety of trematode families. They reported nine species of opecoelid flukes with two-host life cycles, including two species of *Plagioporus* (but not *P. sinitsini*). In contrast, *P. sinitsini* is capable of three-, two-, and one-host life cycles (Dobrovolny, 1939; Barger and Esch, 2000). Lefebvre and Poulin (2005) emphasized throughout their review the importance of the metacercaria wall as a mediator of "...communication to the external world." This is also apparently not the case with *P. sinitsini* because both encysted and unencysted egg-producing adults were observed by Mike inside sporocysts (Barger and Esch, 2000), which is perhaps why they were excluded from the list of Lefebvre and Poulin (2005), which included only species with two-host cycles. Direct transmission to the next host in a one-host life cycle, that is, a snail, could be easily accomplished by the release of whole sporocysts containing eggs. The sporocyst should quickly perish and rapidly decompose, thereby freeing eggs that would hatch after 14 days of maturation.

I thought at the time, and still believe, that the paedogenesis/progenesis phenomenon observed for *P. sinitsini* would have been an interesting line of research for Mike to follow for his PhD but that it was not suitable as a dissertation. My difficulty was that I was not sure where it would take Mike or how long it might require him to generate solid data beyond what he had already produced. I have always felt that it was important for a Master's or PhD student to have a well-defined research problem, one that would allow for the generation of a solid database within a reasonable period of time. Whereas Mike was slightly sidetracked (but productively) at the beginning, his real dissertation work was to focus on the component community of helminth parasites in rosyside dace and redlip shiners.

For some reason, and I have no explanation, much of the parasitology research in aquatic ecosystems up to the time that Mike began his work in the Yadkin River drainage basin had been accomplished in lentic (still water) habitats. The lotic (flowing water) system in which he was to center his work was in part selected for this very reason. If reasonably species-rich component parasite communities in fish could be identified, then we both felt this line of research could be followed effectively. As mentioned previously, he found the component community in dace to have 11 species and 6 in the shiners, and we believed this would be a satisfactory number with which to work.

Hartvigsen and Kennedy (1994) predicted that low-order stream systems like the headwater streams of the Yadkin River should exhibit a linear pattern of similarity in their parasite assemblages. Mike's dataset strongly supported their hypothesis, that is, "...parasite community similarity was strongly related to the distance between sampling sites" (Barger, 2001). The continuum in the component community similarity (as might be expected) was broken at two sites, a dam and a waterfall. It is also of interest that rosyside dace do not appear above the waterfall, but *E. symmetrica* does, and so does *P. sinitsini*, providing very strong evidence to support the notion of a one-host life cycle for the parasite. The waterfall is a barrier for three species of fish, that is, rosyside dace, redlip shiners, and creek chub. Since these three species carry most of the parasite species, it is not surprising that the waterfall served as a break for transmission of the parasites as well. Another interesting aspect of the one-host life cycle is tied to a study conducted by Huryn and Denny (1977). They examined the movement patterns among four species of *Elimia* and found that some populations were quite sedentary, while others were capable of moving up to 40 m in a single month. Moreover, all of the significant movement was upstream, which would also account for the presence of *P. sinitsini* above the waterfall, without dace.

The dam was also involved in separating helminth species longitudinally. Infracommunities in dace below the dam were less species rich than above, while redlip shiners retained about the same parasite richness below the dam as above it. Mike proposed that "...the pattern of community similarity is a result of processes differentially affecting individual component communities" and that this is an "...additive effect..." (Barger, 2001).

Mike found that habitat heterogeneity was "...an important structuring mechanism for the parasite community in Basin Creek" (Barger, 2001). It follows that habitat similarity should parallel similarity in the parasite community. He envisioned the presence of a fish and parasite gradient to match certain environmental gradients, for example, depth of a given habitat. He proposed that the gradient should be coupled with increased species richness in both fish and parasite communities, producing what he referred to as a "...nested arrangement of species." This led him to conclude that in a nested community, "The most species-poor samples will include only those taxa that are ubiquitously distributed in the landscape, whereas species-rich samples will include those taxa and others that are restricted in their distribution" (Atmar and Patterson, 1993). When he completed this part of his research, he concluded that "...each host-parasite combination represents a potentially unique set of ecological parameters that interact with the spatial heterogeneity available, just as each fish is potentially unique in its preferences among habitats and microhabitats" (Barger, 2001).

Overall, Mike identified 22 species of parasites (excluding protozoans) in 18 fish species from among 14 collecting sites in seven streams comprising the Yadkin River headwaters. Studies of community structure generally reveal a positive relationship between the distribution of taxa and their local abundance, that is, commonly referred to as the distribution–abundance relationship (Hanski et al., 1993). Brown (1984) suggested the distribution–abundance relationship concept is based on "...differing abilities of species to utilize the array of resources available to them across a landscape" (Barger and Esch, 2002). In other words, generalist species should exhibit a wide distribution in space and be locally abundant, while specialists should have narrow distributions and not be locally abundant. Mike attempted to test this hypothesis using the data obtained from sampling in the Yadkin headwaters.

The initial point made by Barger and Esch (2002) was that host specificity by the parasite is "... the inverse of resource breadth." If this crucial assumption is correct, then Mike's data supported Brown's (1984) hypothesis regarding the assumed outcome of the distribution–breadth relationship. Barger (2001) then decided to remove the effects of host distribution statistically. The results were not changed, that is, host specificity is still a consistent predictor for the regional distribution of parasite species, while host specificity is likewise a powerful predictor for local abundance. These findings supported both the first and second features of Brown's (1984) hypothesis regarding the distribution–abundance hypothesis.

Another segment of Barger's (2001) dissertation work considered the possibility that fish hosts in mountain streams appear to have relatively narrow home ranges compared with downstream fishes (Gorman, 1986). Although Mike did not rule out this option, he suggested that dispersal would probably play a more significant role in structuring parasite communities in streams with less habitat variability, that is, unstructured streams.

Another graduate student, Joel Fellis, followed up on some of Mike's findings. Joel completed his undergraduate work at the University of New Mexico in the spring of 1998 and matriculated at Wake Forest University in the fall of the same year. He completed his Master's degree in 2001 and his PhD in 2005. His MS research was focused on the component community structure and seasonal dynamics of helminth parasites in green and bluegill sunfishes in Charlie's Pond (see Chapters 6 and 7). Interestingly, the two fish species possessed the same suite of parasites, but when abundance patterns were examined, the parasite component communities in the two species were distinct.

Joel's dissertation research was in lentic habitats, not in streams. Another difference was that Joel examined component communities only in bluegills and simultaneously compared the autogenic and allogenic constituents in each component community for similarity/dissimilarity. He took his fish samples from 10 isolated ponds in the Piedmont area of North Carolina near Charlie's Pond. Joel discovered that the autogenic–allogenic condition had a significant impact on specific features of community structure, including species–surface relationship of the ponds and "…the influence of geographic distance on community dissimilarity" (Fellis, 2005). The first observation, regarding the species–area relationship, was found to apply to allogenic species, but not to autogenic species. This result is not surprising because the colonization abilities of allogenic parasite species are considerably greater than those for autogenic taxa.

The allogenic–autogenic status was an important factor when considering geographic distance and community dissimilarity, just as it was when considering the component community structure in lotic habitats as observed by Mike Barger in the Yadkin headwaters. Joel found that interpond autogenic community dissimilarity was random with respect to interpond distance, but not for allogenic species. This is what one might expect considering that the colonization capability of autogenic species is significantly less than for allogenic species.

With these results in hand, Joel decided to examine what can be defined as community similarity decay among parasite communities in bluegills from ponds located at considerably greater distances from one another (up to 1000 km apart rather than a few) (Fellis and Esch, 2005). Unfortunately, this effort did not pay full dividends when it came to predictions formulated at the outset of the work. The original hypothesis was that the much greater dispersal characteristics of allogenic species should exhibit lower levels of distance decay as compared to autogenic species. When autogenic and allogenic species in a single pond were compared with total community similarity in another pond a great distance away (many kilometers), 39% of the variance was attributed to interpond distance. As Joel stated (Fellis, 2005), "Finding that allogenic community similarity decayed at a greater rate than autogenic similarity was unexpected." He noted that Poulin (2003) observed comparable results when he examined parasite communities in perch and pike in terms of distance decay, but when he compared the data obtained from mammalian hosts, there was an inconclusive pattern. The only explanation for this phenomenon is that he found autogenic communities to be "...less similar to each other than allogenic communities are to each other ($P = 0.021$)" (Fellis, 2005). He reported that while there was a significant distance effect for autogenic species, just a small part (14%) could be explained by distance.

The work accomplished by Barger and Fellis is significant. It reveals a strong resilience in the component parasite community structure in widely separated piscine taxa inhabiting both flowing and still water aquatic systems. Both students generated substantive datasets regarding comparative parasite community structure in lentic and lotic habitats.

References

Atmar, W., and B.D. Patterson. 1993. The measure of order and disorder in the distribution of species in a fragmented habitat. *Oecologia* **96**: 373–382.

Barger, M.A. 2001. Landscape ecology and the influence of fish and parasite communities of streams in the Appalachian Mountains of North Carolina. Ph.D. dissertation. Wake Forest University, Winston-Salem, North Carolina, 266 p.

Barger, M.A., and G.W. Esch. 1999. *Allopodocotyle chiliticorum* n. sp. (Digenea: Opecoelidae) from redlip shiners, *Notropis chilticus*, in Basin Creek, North Carolina. *Journal of Parasitology* **85**: 891–892.

Barger, M.A., and G.W. Esch. 2000. *Plagioporus sinitsini* (Digenea: Opecoelidae): A one-host life cycle. *Journal of Parasitology* **86**: 150–153.

Barger, M.A., and G.W. Esch. 2002. Host specificity and the distribution-abundance relationship in a community of parasites infecting fishes in streams in North Carolina. *Journal of Parasitology* **88**: 446–453.

Brooks, D.R., and D.A. McLennan. 1993. Historical ecology: Examining phylogenetic components of community evolution. *In* Species diversity in ecological communities: Historical and geographical perspectives, R.E. Ricklefs, and D. Schluter (eds.). University of Chicago Press, Chicago, IL, p. 267–280.

Brown, J.H. 1984. On the relationship between diversity and abundance. *American Naturalist* **124**: 255–279.

Dobrovolny, C.G. 1939. Life history of *Plagioporus sinitsini* Mueller and embryology of new cotylocercus cercariae (Trematoda). *Transactions of the American Microscopical Society* **58**: 121–155.

Fellis, K.J. 2005. Effects of life cycle variation on community assembly and population genetic structure of macroparasites in the bluegill sunfish, *Lepomis macrochirus*. Ph.D. dissertation. Wake Forest University, Winston-Salem, North Carolina, 85 p.

Fellis, K.J., and G.W. Esch. 2005. Variation in life cycle affects the distance decay of similarity among bluegill sunfish parasite communities. *Journal of Parasitology* **91**: 1485–1486.

Gorman, O.T. 1986. Assemblage organization of stream fishes: The effects of rivers on adventitious streams. *American Naturalist* **128**: 611–616.

Gould, S.J. 1977. Ontogeny and phylogeny. Belnap Press of Harvard University Press, Cambridge, MA, 495 p.

Hanski, I., J. Kouki, and A. Halkka. 1993. Three explanations of the positive relationship between distribution and abundance of species. *In* Species diversity in ecological communities: Historical and geographical perspectives, R.E. Ricklefs, and D. Schluter (eds.). University of Chicago Press, Chicago, IL, p. 108–116.

Hartvigsen, R., and C.R. Kennedy. 1994. Patterns in the composition and richness of helminth communities in brown trout, *Salmo trutta*, in a group of reservoirs. *Journal of Fish Biology* **43**: 603–615.

Huryn, A.D., and M.W. Denny. 1977. A biomechanical hypothesis explaining upstream movements by the freshwater snail *Elimia*. *Functional Ecology* **11**: 472–483.

Lefebvre, F., and R. Poulin. 2005. Progenesis in digenean trematodes: A taxonomic and synthetic overview of species reproducing in their second intermediate hosts. *Parasitology* **130**: 587–605.

Poulin, R. 2003. The decay of similarity with geographical distance in parasite communities of vertebrate hosts. *Journal of Biogeography* **30**: 1609–1615.

Raf, R. 1996. The shape of life: Genes, development, and the evolution of animal form. University of Chicago Press, Chicago, IL.

11 Red Sore Disease

Progress, therefore, is not an accident, but a necessity.... It is part of nature.
Social Statistics, Herbert Spencer (1820–1903)

For 5 years, between 1975 and 1980, I was a coprincipal investigator on grants totaling almost a half million dollars; my partner was a graduate student by the name of Terry Hazen. Today, this amount of money would seem like "peanuts," but this was approximately 35–40 years ago, when a dollar was worth almost a dollar. In fact, the funding agencies were "chasing" us with offers of even more money. The really ironic part of this was that the research focus was directed at a bacterium, *Aeromonas hydrophila*, the causative agent for something called red sore disease (Figure 11.1) in fish. I say ironic because I am not a microbiologist—in fact, I have never had a course in microbiology. I fell into the situation quite by accident.

During the early 1970s in the southeastern part of the United States, and especially North Carolina, a series of red sore epizootics occurred in several lakes and estuaries. Thousands of game and commercial fish turned "belly up" over very short periods of time. One of the first of these occurrences took place in Lake Apopka, Florida, in 1971, where, in short order, approximately 120,000 fish were killed, as well as numerous snakes, turtles, and alligators in the same reservoir. Among the piscine species were largemouth bass, other centrarchids, and striped bass. In our part of the world, whenever largemouth bass are affected negatively, one can be certain that something will "hit the fan," which in large part accounts for the financial support we were to acquire so easily for our red sore work. In Badin Lake, North Carolina, in 1973, an estimated 37,500 fish died from red sore disease over a period of 13 days. During the fall of 1976, approximately 95% of the white perch population was killed in Albemarle Sound, North Carolina; in the same outbreak, nearly 50% of

Ecological Parasitology: Reflections on 50 Years of Research in Aquatic Ecosystems,
First Edition. Gerald W. Esch.
© 2016 John Wiley & Sons, Ltd. Published 2016 by John Wiley & Sons, Ltd.

Figure 11.1   Lesions on the surface of *Micropterus salmoides*, caused by *Aeromonas hydrophila*.

the commercial fish caught in the Sound had to be discarded because of unsightly surface lesions associated with red sore disease.

All of these fish were killed by *A. hydrophila* in combination with a peritrich ciliate, *Epistylis* spp., which, it was originally thought, together caused red sore disease. The initial scenario suggested that a motile telotroch stage of the protozoan attached to the surface of the host and produced a stalk with feeding bodies that caused scale erosion. Via these surface lesions, *A. hydrophila* was thought to enter subepidermal surface tissues and eventually induce hemorrhagic septicemia, selective organ destruction, and host death.

It was about this time, in 1973, that I met my partner on this project. Terry Hazen was a big man on a big frame who came down from Michigan State University (MSU) in East Lansing to the Kellogg Biological Station (KBS) at Gull Lake to take my course in field parasitology. When I spoke with him while writing this essay, he reminded me that his particular MSU graduate program required him to take courses in every biological department at MSU, including microbiology. Since parasitology was taught in that department, he decided a summer at Gull Lake would be of interest to him. While a formal thesis was not required, a research program was, which led him to Gull Lake and work on amphipods and *Crepidostomum cooperi*, the allocreadiid fluke, which was described in Chapter 4. Most important for me, however, Terry had a solid background in microbiology, just what was needed for the red sore project.

The summer of 1974 would be my last at Gull Lake. For a while, I also thought it would be my last year at Wake Forest University (we even sold our house in Winston-Salem and bought another in Aiken, South Carolina, where we were to live for the next year). I had been offered an opportunity to go to the Savannah River Ecology

Laboratory (SREL) with a staff appointment, plus a faculty position at the University of Georgia. Initially, I had decided to resign from Wake Forest and leave permanently. However, my wife, Ann, while shopping for groceries 1 day, accidentally encountered Emily Wilson, the wife of our Provost at Wake Forest, Ed Wilson. They engaged in a conversation during which Ann told Emily that I was leaving Wake Forest to take the job at SREL. The next day, I received a telephone call from Ed who invited me over to visit with him, which I did. He persuaded me to not resign, but to take a sabbatical instead of "burning the Wake Forest bridge." I followed his advice, and eventually, everything turned out very well. The spring I left Wake Forest, Herman Eure had completed his PhD and was offered my job at Wake Forest while I was on leave at SREL. When I came back as the new chair of the department in 1975, we offered Herman a tenure-track job and he took it. So, Herman and I both wound up teaching at Wake Forest. It was a huge break for each of us.

All staff members at SREL had their own technicians. I knew about the presence of the red sore problem in Par Pond (see Chapter 4) and had by then decided that the disease was going to be my focus during my leave at SREL. Since I was not a microbiologist, I obviously needed one. So, I offered Terry the job with the following arrangement. He was to come down to SREL as my technician for a year (the academic year of 1974–1975). If I returned to Wake Forest, he would come back with me as my PhD student; if I stayed at SREL and the University of Georgia, he would matriculate there as my PhD student. We would work together as equal partners, with him focusing on the microbiological/ecological aspects of *A. hydrophila*, and I would concentrate more on the epizootiological/physiological features of the disease in fish. We agreed that this was also the way we would publish any papers that would emerge as a result of our work. During the course of our research on red sore disease, we also asked a couple of specialists to collaborate with us. Things could not have worked out better for both Terry and me and for our collaborators along the way!

One of the first things that we wanted to accomplish was to confirm the supposed role for *Epistylis* sp. in the disease process. Unfortunately, or fortunately, depending on how one considers the situation, Jiri Lom, the outstanding Czech protozoan parasitologist, beat us "to the punch." In a paper Lom (1973) wrote, he suggested that there was no evidence that *Epistylis* sp. was capable of producing histolytic enzymes necessary to cause scale erosion. Terry and I held the same view, but we believed there was a need for us to confirm or reject Lom's hypothesis. So, using a combination of electron microscopy, plus immunofluorescent and culturing techniques, we developed sufficient evidence to conclude that *Epistylis* sp. had no role in red sore disease, thereby supporting Jiri Lom's idea. The stalk of the peritich protozoan, which is the attachment device to the fish surface, does not possess the internal organelles necessary to produce histolytic enzymes. Moreover, there was no indication that the protozoan was involved in producing scale erosion of the kind seen in red sore disease. Of more than 100 lesions carefully examined, we were able to isolate *A. hydrophila* more than 95% of the time, but *Epistylis* sp. was associated with the lesions less than 30% of the time. In addition, we were able to experimentally induce surface lesions in bass by direct exposure to *A. hydrophila* in the water, in the complete absence of *Epistylis* sp. Finally, Hazen et al. (1979) were also able to induce red sore lesions on alligators in the absence of *Epistylis* sp.

As an aside at this point, Terry (Esch et al., 1976) had developed an elegant technique for fixing and staining epidermal tissues associated with *Epistylis* sp. and *A. hydrophila* using phosphate buffered formalin as a fixative and Semicohn's acetocarmine as the stain. Acetocarmine is typically employed for whole mounts of tapeworms and trematodes, not protozoans, but the results he produced were exceptional. In fact, he took the slides to a well-respected protozoologist at Georgetown University in Washington, DC, who was greatly impressed regarding their quality and even asked Terry about his histological technique because he was in such awe of the outcome. He was enthusiastically encouraged by our Georgetown colleague to get the work published as soon as possible. So, when Terry returned from Washington, DC, he put the paper together and sent it to the *Transactions of the American Microscopical Society* (TAMS; now called *Invertebrate Biology*); it was accepted immediately by return mail. In fact, after publication of the paper, he quickly ran out of reprints (yes, reprints—remember, this was in the good old days of "paper"). Not very long after the work was published, Terry was contacted by the editor of *Microscopica Acta* and invited to submit the Hazen et al. (1976) TAMS paper for republication in his journal—not too shabby! Terry contacted the TAMS editor who supported the idea if he would include a statement that the paper was first published in TAMS.

The external lesions on the surface of fish with red sore disease are quite distinct. Since fish were dying in association with the external lesions, we also surmised that organ system physiology was altered in some way by the presence of systemic *A. hydrophila*. In other words, there was a clear need for a thorough histopathology investigation. At this point, we asked Harry Huizinga to join us in the study. An old friend, Harry was an excellent histologist and fish pathologist and accepted our invitation (Huizinga et al., 1979). We observed that the epidermis covering the scales was completely eliminated in the surface lesions. The adjacent dermal tissue would become hyperplastic, accompanied by intense mononuclear infiltration, hemorrhage, and necrosis. The underlying muscle layers were edematous, with severe inflammation and necrosis also evident. In more serious and extensive infections, especially those associated with the dorsal fins and tails, necrosis was so complete that denuding of tissue extended all the way to the bone substructure, with functional morphology most likely destroyed permanently. Internal organs affected included the heart, liver, spleen, and kidney; the latter organ was especially damaged. In the posterior kidneys of some bass, Bowman's capsules and renal tubules were completely destroyed; histologically, the kidneys looked like someone had mashed them with a boot. Liu (1961) described the production of what he termed, "potent exotoxins," by *A. hydrophila* that were capable of causing external lesions and internal tissue destruction of the kind we were seeing in largemouth bass.

It is of interest to note that the external histopathology associated with red sore disease appeared to vary with different host species. Among bluegill sunfish, surface lesions were most frequently seen on the dorsal and anal fins (Esch et al., 1976). On largemouth and striped bass, the lesions occurred with greatest regularity along the lateral line, ranging from the gill plates to the tail. For channel catfish, lesions were most common on the head and mouth. The variability of lesion location according to host species raised the possibility of strain differences in *A. hydrophila*, but we did not pursue this line of work. There appeared to be some variability of red sore disease relative to the host species in Par Pond as well. Prevalence was as high as 75% in largemouth bass, but zero among black crappie (*Pomoxis nigromaculatus*), another centrarchid species. Terry told me recently that a number of new species of *Aeromonas* had been described between then and the present time, some of which were more virulent than the one with which we were working back in the 1970s. However, it is of interest to note that the frequency of massive fish kills reported in the 1970s has abated and, for what reason, we are unsure. Terry speculated, however, it is possible that periodic outbreaks still occur, but that they do not warrant the same level of publicity they did when first observed back in the 1970s.

Before moving to other issues regarding the biology of red sore disease in fish, several aspects of the ecology of *A. hydrophila* require a description. Terry conducted this research (Hazen, 1978, 1983) largely as part of his dissertation. Most of this effort was undertaken in Par Pond, the sizeable cooling reservoir located on the SRP near Aiken, South Carolina (as described in Chapter 3). Densities of *A. hydrophila* were measured at several heated and ambient temperature sites. A wide range of physical and chemical parameters were measured in the reservoir, including temperature, dissolved oxygen, pH, conductivity, redox potential, total organic carbon, inorganic carbon, particulate organic carbon, and dissolved organic carbon. Without going into great detail, it can be said that *A. hydrophila* was most abundant in heated parts of the reservoir, but could not be isolated at depths greater than 1 cm from the water surface. Highest densities were present in surface microlayers associated with transient organic slicks and decomposing Asian milfoil (*Myriophyllum spicatum*). Seasonally, the highest densities occurred in the spring months of March through May, with a second peak in the fall, corresponding to the fall mixis. In Par Pond, the most consistent physical parameter relative to *A. hydrophila* density was temperature. If temperatures are elevated, *A. hydrophila* density was as well. If *A. hydrophila* density was high, there either was, or would be, an elevated probability of red sore disease.

Terry was also interested in a wider geographic distribution of *A. hydrophila* (Hazen et al., 1978). He sampled 147 lentic and lotic sites in 30 different states. Only 12 of these localities were devoid of the bacteria. He referred to these locations as "extreme," for example, hypersaline lakes, geothermal springs, heavily polluted (primarily industrial waste) rivers, etc.

In the years between 1968 and 1977, more than 10,000 largemouth bass were taken from Par Pond by electrofishing, seining, gill netting, and ordinary hook and line (Gibbons et al., 1978). The weight–length relationship for each bass was determined, and body condition, or K-factor, of each fish was calculated. When K-factors for all bass were examined seasonally, there was a clear pattern. Maximum body condition was observed during winter (December, January, February), followed by a slight decline in spring (March, April, May), then a large drop in summer (June, July, August), and recovery in fall (September, October, November). Occasionally, the lowest K-factors were seen in the fall. The decline from spring to summer was a consequence of postspawning, as might be expected. However, we also asserted that thermal effluent from the nuclear production was an important cause of deteriorating body condition as well (Esch and Hazen, 1978).

Body condition, or K-factor, is an important measure of physical fitness. Overall, fish present in areas of Par Pond impacted by thermal effluent exhibited lower K-factors. During 1974–1975, Terry, an undergraduate student (Tom Quinn), and I collected several hundred largemouth bass from the heated arm of the reservoir (Quinn et al., 1978). After obtaining the measurements used for calculating body condition, these fish were tagged and released at a central point in the heated arm of Par Pond. Subsequently, a substantial number were recaptured, enough to conclude that many bass appeared to be returning to the same locality in which they were first caught. In other words, the thermal effluent was neither forcing them out of the heated arm nor was it attracting fish from ambient temperature localities.

Because of within season variability in body condition and prevalence of red sore disease, we conjectured that there could be a relationship between the body condition and the probability of a given bass becoming infected with *A. hydrophila*. When infection percentages for each 0.2 unit K-factor subclass between 1.0 and 3.0 were compared, the data clearly revealed that bass with the lowest (poorest) body conditions were the most likely to have *A. hydrophila* lesions. The highest red sore prevalences were in subclasses between 1.0 and 1.8. After the K-factor increased above 2.0, the prevalence of red sore disease rapidly declined. In a supporting study, Gibbons et al. (1978) observed that bass with body conditions below 2.0 were mostly without dissectable body fat, an excellent indicator of poor physical fitness.

Our observations led us to several conclusions, plus a hypothesis regarding the relationship between red sore disease, temperature, and body condition (Esch and Hazen, 1978). We had established that there was a distinct relationship between the body condition and the probability of infection in bass. Moreover, body condition and probability of infection were treated as independent variables, except that the association was muted among bass caught in thermally altered areas of Par Pond.

Based on these findings, we proposed that elevated water temperatures were reducing body condition in conjunction with stress, especially in areas of the reservoir receiving exceptionally warm water from the nuclear production facilities. We suggested (Esch and Hazen, 1978) that as water temperatures increased, metabolic rates were stimulated. Accompanying the increased metabolic activity at higher temperatures, both body fat and protein declined, along with body condition. However, the increased probability of acquiring red sore disease was most likely associated with stress as will be shown subsequently.

Hans Selye, a Hungarian immigrant who was a faculty member at the Universite de Montreal, developed major components of the stress concept. His assertion was that a stress response occurred as a three-step series of reactions he called the General Adaptation Syndrome (GAS). In sequence, the GAS included an alarm reaction, a resistance stage, and a period of exhaustion. While we agree that stress at the individual level can be viewed in this manner, we felt that Selye's idea did not extend to the population or ecosystem levels and that the exclusion of plants was also an oversight (Esch et al., 1975). We revised Selye's definition by saying that stress is the affect of any force that tends to extend any homeostatic or stabilizing process beyond its normal limit, at any level of biological organization. Stress can be identified as the product of change in homeostatic or environmental instability.

In vertebrate animals, a result of stress is the production and release of excess quantities of corticosteroids, such as cortisone or cortisol. Both of these steroids are produced in the cortical region of the adrenal gland, but cortisol is the primary stress hormone in fish. It has multiple functions, one of which is to suppress the immune system; another role is to mitigate damage to cell membranes during periods of stress. An insidious and clearly undesirable side effect is the concomitant reduction in the circulation of leukocytes (leukopenia), which results in increased vulnerability to infection by pathogenic agents.

As previously indicated, we proposed (Esch and Hazen, 1978) that red sore disease was related to stress stimulated by elevated water temperature. It was our proposition that one of the side effects of the thermal effluent was elevation of metabolism that produced declines in body condition in largemouth bass of Par Pond. Simultaneous with thermal stress was increased cortisol production, which led to leukopenia and greater susceptibility to infection by *A. hydrophila*. It should be noted that this sort of scenario had been suggested by other investigators of the time but rarely tested. We, therefore, undertook a special study to assess the validity of our hypothesis in a natural population of largemouth bass in Par Pond.

During a period from the summer of 1976 and extending through the winter of 1978, we examined 465 largemouth bass. All of these fish were weighed, measured, and bled. Using weight and length, we computed the K-factors for each fish. Five blood parameters were checked from 397 bass, that is, hematocrit, hemoglobin, total red blood cells, total white blood cells, and cortisol. The fish were divided into two groups, those with K-factors less than 2.0 and those greater than 2.0. The results were as we predicted. The mean values for each blood parameter were significantly different in bass with K-factors less than 2.0 and greater than 2.0. We were concerned that division of the bass into two subclasses could have produced a "quantitative artifact." So, we pooled the bass data and evaluated them using a multiple regression analysis, with K as the dependent variable. The correlation matrix indicated significant relationships between each blood parameter and body condition.

Based on these results, we proposed that as stored energy reserves in bass are reduced and that body condition, or physical fitness, declines. Simultaneously, hematocrit, hemoglobin, and total red blood cell counts also drop. Most importantly, cortisol levels increase and white blood cells decrease. Consequently, vulnerability to infection by *A. hydrophila* should increase. We do not know, however, if leukopenia is related to suppression of antibody production or the lysis of phagocytes.

Before continuing with the stress story, I want to insert information regarding a part of our work in Albemarle Sound, a large estuary situated in the northeastern part of North Carolina. The Sound is created by the confluence of the Roanoke and Chowan Rivers. The latter stream is formed by joining the Meherrin, Nottaway, and Blackwater Rivers, which, along with the Roanoke, have their headwaters in Virginia. I make note of that here because much of the negative aspects of the water conditions in the Sound at the time of our study could be sourced to industrial pollution from both Virginia and North Carolina. To put it bluntly, Albemarle Sound was a real mess.

The water parameters examined by Terry during the work in Albemarle Sound were very similar to those performed in Par Pond. Recall that in Par Pond when water temperatures were high, we also saw high densities of *A. hydrophila*. In the Sound, the highest densities of *A. hydrophila* were associated with elevated phosphate and total phosphorus. Wetzel (1975) stated that in unpolluted surface water, total phosphorus concentrations should range from 10 to 50 μg/L. At 13 collecting sites in Albemarle Sound, 292 water samples were examined between April 1977 and July 1979. Collections from the Sound exceeded 50 μg/L a total of 254 times or in about 85% of the samples total phosphorus exceeded what we would expect in unpolluted water. Thus, most of the time, sampling stations with elevated phosphorus could be considered as eutrophic, or even hypereutrophic. However, care must be taken in drawing conclusions from these data because we were dealing with correlations only, not direct cause and effect relationships in Albemarle Sound, or even Par Pond for that matter. Nonetheless, the data are formidable and, when viewed in the context of stress as a factor in red sore disease, they are very reasonable.

An important question for us at that point in time was related to the differences in red sore disease biology in Par Pond versus Albemarle Sound. In the reservoir, we could easily relate the disease to thermal effluent and stress. However, heated effluent was not an issue in Albemarle Sound. Moreover, red sore disease in Par Pond appeared to be a chronic phenomenon, while in Albemarle Sound, the disease was associated with a sudden and massive fish kill involving the white perch population, although we also knew that at least half of commercially caught fish had to be discarded because of ugly surface lesions. In other words, the epizootiological characteristics of the disease were quite different in the two aquatic ecosystems.

An important facet of the study revealed that several water quality parameters are associated with increased densities of *A. hydrophila*, which, in turn, can be associated with epizootic outbreaks of red sore disease. Just as significant, several of these same water quality parameters, if altered substantially, may also promote stress. For example, it is well known that increased nitrogen, organic carbon, and/or phosphorus loading, along with higher summer water temperature, can lead to lower dissolved oxygen and acute stress in some fish species. With stress, there is increased susceptibility to infection, which can lead to sudden and massive fish mortality such as the kind observed in Albemarle Sound.

A short anecdote should be inserted at this point. I become very nervous when I fly, no matter the type of airplane, or the distance traveled, or the airline, etc. While doing the work in Albemarle Sound, Terry decided he needed to learn how to fly (I am not sure why, but he probably was influenced by a pilot colleague [Ray Kuhn] who I referred to as a research partner of mine in Chapter 1). Accordingly, Terry took lessons and secured a pilot's license after completing all of the requirements.

During our work, we decided it would be worth our while to develop personal relationships with Bassmasters, an international organization devoted to largemouth bass fishing, conservation, etc. It was our hope that they would help in identifying bodies of water in which red sore disease might be a problem so that we could extend our research in a more productive manner. Terry had heard about a Bassmasters club meeting over in the eastern part of North Carolina and decided to attend so we could introduce ourselves. He asked them if we could make a presentation to describe our work and they invited us to "come on over."

Terry asked me to fly with him, but I quickly said no. So, Terry flew over and back alone. However, he was confronted by two problems during his return. First, he ran out of gas at about 3000 ft up in the air just a few miles from the airport at which he was to land (he claims that he stopped in Raleigh and topped out his fuel tank, but that is questionable). Second, he did not have the altitude necessary to reach the airport. However, two good things also emerged during his crisis. First, he found a straight country road on which to land (he was very lucky in this regard because a straight country road in North Carolina is most unusual). Second, he was not seriously hurt (although the plane was totaled)! I continue to fly, but not with great relish. Terry does too, but now on big jet planes.

As noted earlier, massive red sore disease outbreaks have not been reported in recent years. Not long after the red sore work in Albemarle Sound and Par Pond was completed in 1978, I am happy to say that my microbiology "research" came to a halt. Terry continued with his microbiology and has since become a renowned microbial ecologist. After stints at the University of Puerto Rico, several years back at the Savannah River Plant, and the Lawrence Laboratory in Berkeley, California, he was recently appointed to a prestigious Governor's Chair at the University of Tennessee in Knoxville, where he now resides and works.

References

Esch, G.W., and T.C. Hazen. 1978. Thermal ecology and stress: A case history for red-sore disease in bass. *In* Energy and environmental stress in aquatic ecosystems, G.W. Esch, and R. McFarlane (eds.). ERDA Symposium Series, (CONF-77114). National Technical Information Service, U.S. Department of Commerce, Springfield, VA, p. 331–363.

Esch, G.W., J.W. Gibbons, and J.E. Bourque. 1975. An analysis of the relationship between stress and parasitism. *American Midland Naturalist* **93**: 539–553.

Esch, G.W., T.C. Hazen, R.V. Dimock, Jr., and J.W. Gibbons. 1976. Thermal effluent and the epizootiology of the ciliate *Epistylis* and the bacterium *Aeromonas* in association with centrarchid fish. *Transactions of the American Microscopical Society* **95**: 687–693.

Gibbons, J.W., D.E. Bennett, G.W. Esch, and T.C. Hazen. 1978. Relationship of thermal effluent and body condition of largemouth bass in a South Carolina cooling. *Nature (London)* **274**: 470–471.

Hazen, T.C. 1978. The ecology of *Aeromonas hydrophila* in a South Carolina cooling reservoir. Ph.D. dissertation. Wake Forest University, Winston-Salem, North Carolina, 259 p.

Hazen, T.C. 1983. A model for the density of *Aeromonas hydrophila* in Albemarle Sound, North Carolina. *Microbial Ecology* **9**: 137–153.

Hazen, T.C., G. Smith, and R.V. Dimock, Jr. 1976. A method for fixing and staining peritrich ciliates. *Transactions of the American Microscopical Society* **95**: 687–693.

Hazen, T.C., M.C. Raker, G.W. Esch, and C.P. Fliermans. 1978. Ultrastructure of red-sore lesions on largemouth bass (*Micropterus salmoides*). *Journal of Protozoology* **25**: 351–355.

Hazen, T.C., R.V. Gordon, C.P. Fliermans, and G.W. Esch. 1979. Isolation of *Aeromonas hydrophila* from the American alligator, *Alligator mississippiensis*. *Journal of Wildlife Diseases* **15**: 239–243.

Huizinga, H., G.W. Esch, and T.C. Hazen. 1979. Histopathology of red-sore disease in largemouth bass, *Micropterus salmoides*. *Journal of Fish Diseases* **2**: 310–321.

Liu, P.V. 1961. Observations of the specificities of extra-cellular antigens of the genera *Aeromonas* and *Serratia*. *Journal of General Microbiology* **24**: 145–153.

Lom, J. 1973. Sessiline peritrichs on the surface of some freshwater fishes. *Folia Parasitologica* **1**: 36–56.

Quinn, T., G.W. Esch, T.C. Hazen, and J.W. Gibbons. 1978. Long range movement of largemouth bass (*Micropterus salmoides*) in a thermally altered reservoir. *Copeia* **1978**: 542–545.

Wetzel, R.G. 1975. Limnology. Saunders, Philadelphia, PA, 489 p.

12 The End, Almost

Nature is a mutable cloud, which is always and never the same.
Essays: First Series [1841]. *History*, Ralph Waldo Emerson (1803–1882)

Not very often does one have the opportunity of teaching the child of a former student; however, I was offered that chance when Lauren Camp was awarded a teaching assistantship in our department in 2005. Her father, Joe Camp, was a PhD student of mine who completed his dissertation in August 1980. Lauren was unusual in another respect as well. When she arrived here to begin work on her master's degree, she had not taken a course in parasitology. So, it was left for me to teach her the basics of our discipline and then to acquaint her with field research. I was very pleased to have her as a student because she was to eventually expose Kyle Luth, Mike Zimmermann, and me to an entirely new model (for us) in the study of host–parasite relationships.

For a great many years, my students and I focused on the biology of snail–trematode systems, mostly in Charlie's Pond, which is about 18 miles northeast of Winston-Salem and our university's campus. Nick Negovetich (see Chapter 7) was the first of my graduate students to discover a rather odd host–parasite system in another pond located about 6 miles to the southwest of us. Just outside our city is Tanglewood Park, complete with a couple of golf courses and several other amenities. Adjacent to one of the golf courses is Mallard Lake, an impoundment, slightly larger than Charlie's Pond.

Nick had collected snails from several ponds, including Mallard Lake, in the vicinity of Winston-Salem in the fall of 2003 and spring of 2004. He was interested in a large-scale comparison of the parasite communities between local ponds and how snail communities (differences in presence/absence and abundance) might predict a pattern of parasite community structure. Joel Fellis, another of my PhD

Ecological Parasitology: Reflections on 50 Years of Research in Aquatic Ecosystems,
First Edition. Gerald W. Esch.
© 2016 John Wiley & Sons, Ltd. Published 2016 by John Wiley & Sons, Ltd.

students, had surveyed the parasite community in the bluegill/green sunfish from Mallard Lake to identify the autogenic/allogenic trematode species but had not examined any snails for cercariae being shed. When Nick first sampled there, he observed several types of cercariae, including two or three different echinostome species being shed from *Helisoma (Hel.) anceps*. At some point during the spring collection, he also began to see the nematodes in flesh of the snails. In a recent telephone conversation, he recalled for me that nearly every snail possessed one, or more, specimens of the same nematode. He told me that based on previous experience, he was not surprised to find nematodes while crushing and dissecting the snails. He said that he regularly picked up free-living nematodes with snails and at first thought these new worms were simply "hitching" a ride in the spire or umbilicus depression. They could have also been carried with the leaf litter or pond muck that is usually collected along with the snails. Occasionally, he found a nematode in the digestive tract of snails. His recollection was that he at first thought the rare intestinal nematodes were actually free-living species, which were accidentally consumed by the snails. However, as he began to see the same nematodes in nearly every snail collected from Mallard Lake, he began to wonder if these nematodes were using the snails as a host of some sort.

He recalled casually mentioning to Joel Fellis one morning that he was finding all of these nematodes in *Hel. anceps* and that Joel quickly examined them using a compound scope to see if they were larvae or adults. His observations prompted him to seek advice from me. Nick vaguely remembers that I was at first skeptical about the nematodes actually being parasitic within the snails (I must admit that, at the time, I knew nothing about snail nematodes). During my career, I had necropsied a lot of snails, but I do not recall encountering even one with a nematode. However, when Joel and Nick reminded me that nearly every snail was infected with the same worms, I was won over to their conclusion regarding the presence of parasitic nematodes in snails from Mallard Lake.

While Joel was quite excited by the finding of what was quickly identified as *Daubaylia potomaca*, a rhabditid nematode in *Hel. anceps*, he was almost finished with his dissertation research (in 2005) and would not be working with the snail/nematode system after leaving Wake Forest. Nick had very early focused his attention on *Halipegus occidualis* in Charlie's Pond and decided to work in that direction because of its potential for mathematical modeling. Accordingly, when Lauren arrived in 2005, I thought *D. potomaca* and *Hel. anceps* would be ideal to use in pursuit of her master's degree.

While the parasitic nematode was new to us, the students discovered it had been first described from the Potomac River in Virginia by Chitwood and Chitwood (1934). Later, it was reported in *Helisoma trivolvis* and *Helisoma antrosa* from Douglas Lake, Michigan (Chernin et al., 1960). In a personal communication, Cheryl Davis (a great friend of mine and a graduate student of Ray Kuhn) told us that she had found *D. potomaca* in a golf course pond in Norman, Oklahoma, in the

1980s. While recently attending the American Society of Parasitologists annual meeting (2012) in Richmond, Virginia, Kyle Luth was out jogging along the James River and found some *Hel. anceps* in shallow water along the shoreline. He picked up several and carted them home to Winston-Salem where Mike Zimmermann necropsied them, and, voila, he found *D. potomaca*! Finally, not long ago, the same nematode was found by Lauren in a pond on the campus of the University of California–Davis, and another friend, Sam Loker, reported that he had recently found it in New Mexico. While the parasite thus appears to have a very extensive geographic distribution in the United States, we have absolutely no idea how such a wide dispersal was (is) accomplished. In this regard, there were several other golf course ponds close by Mallard Lake at Tanglewood Park with a similar molluscan fauna, but *D. potomaca* is not present in any of them.

In the literature, Lauren found a couple of reports from Eli Chernin and his colleagues (Chernin et al., 1960; Chernin, 1962) that were focused on the possibility of using *D. potomaca* from the University of Michigan's field station at Douglas Lake as a biological control agent for *Biomphalaria glabrata*, the intermediate host for *Schistosoma mansoni*. Their work lasted for a couple of years but was abandoned when they realized that the nematode was ineffective for this purpose. Nonetheless, during their research, they made some very interesting, and peculiar, observations that were to be useful to Lauren for beginning her master's research on the biology of *D. potomaca* in Mallard Lake.

I should also note here that about the time Lauren finished (in 2007), Kyle Luth and Mike Zimmermann came into my lab in 2008, both pursuing their master's degrees. Immediately after Lauren completed her thesis research, Mike followed up on her results and extended them significantly. The *D. potomaca* master's degree project that Kyle was to work on had to be stopped for technical reasons during his first semester as a graduate student. The study that he developed in its place turned out to be a productive project, and he was able to complete it in time to obtain his MS degree, along with Mike. Like Mike, Kyle also decided to stay with me for his PhD degree. So, while I endeavor to tell Lauren's *D. potomaca* story, I will also attempt to integrate some findings by Mike and Kyle in Mallard Lake. Altogether, their work produces a very interesting research narrative.

Chernin et al. (1960) and Chernin (1962) reported that they observed all stages of *D. potomaca*, from eggs to adults, scattered throughout the tissues and organs of their snail hosts and that, accordingly, there was no site predilection by the parasite. Moreover, there was also no evidence of serious pathology in the snail hosts, although there was some localized tissue destruction due to internal migratory habits of the worms. Interestingly, they also observed that large numbers of *D. potomaca* would exit the snails just prior to host death.

In his work with uninfected, lab-reared *Helisoma* spp., Mike repeated an experiment described by Chernin (1962). He exposed snails to adult female nematodes that had previously emerged from other snails. All of the worms disappeared within 12 hours. One to three days after exposure, the snails were necropsied, and nearly all of the adult females were recovered, indicating that this gender of *D. potomaca* must be an infective stage of the parasite.

Based on these observations, Chernin (1962) concluded and Mike confirmed (Zimmermann, 2010) that *D. potomaca* has a heteroxenous (two-host) life cycle but a very unusual one. The term heteroxenous says that there is more than one obligatory host in the life cycle of the parasite. How does this apply to *D. potomaca*? It was determined that adult females exit a snail host and are then recruited by other snails. Mike never observed a male parasite to leave its snail host. This strongly suggested that males are monoxenous, whereas females are heteroxenous, a really weird situation based on the accepted definitions for the monoxenous and heteroxenous terminology.

With her preliminary work completed, Lauren developed a workable set of goals for her thesis research. Because a great deal of the effort by Chernin and his colleagues was centered on *B. glabrata*, she decided to employ only *Hel. anceps* snails since this is the primary snail host in Mallard Lake (three other snail species in the lake [*Menetus dilatatus*, *Physa acuta*, and *Lymnaea columella*] were uninfected by the nematode). Her effort was, in part, designed to determine if the observations for the *B. glabrata–D. potomaca* system would produce similar results for *D. potomaca* in *Hel. anceps*. Accordingly, her objectives were to (i) examine the distribution of the parasite in individual *Hel. anceps* snails, (ii) determine which life cycle stages were present in the snails, (iii) examine the histopathology in *Hel. anceps*, (iv) confirm that females are the infective stage for snails and that they are the dominant emergent sex, (v) establish the life cycle pattern for *D. potomaca* in Mallard Lake, and (vi) determine the seasonal changes in prevalence and intensity/abundance of the parasite.

Mike's research (Zimmermann, 2010) overlapped that of Lauren in several areas. However, he later refined her approach somewhat by, for example, separating the pond into quadrats and collecting snails as though they were from different ponds. He also made note of the substratum identity from which the snails were removed. In this way, he was applying several of the methodologies typically employed in landscape ecology.

Trematodes in their snail hosts generally exhibit an internal site preference, mostly the hepatopancreas or gonads or both. In contrast, she found that *D. potomaca* showed no site preference in *Hel. anceps*, confirming observations by Chernin (1962) for the nematode in *B. glabrata*. The parasites were scattered in all tissues, along with tiny holes in sectioned tissues, suggesting a random internal migratory pattern with relatively little tissue damage, also similar to the Chernin et al. (1960) observations in *B. glabrata*. Neither Lauren nor Mike nor Chernin, however, was able to determine the route of first entry into the snail. However, all three of them agreed that recruitment of *D. potomaca* had to be either direct (through the foot) or that the nematode could be ingested (eaten). (I was reminded by Kyle Luth that

Hel. anceps is a nonoperculate snail and that *D. potomaca* would likely have easy access at any time to any of the exposed tissue, inside or outside the shell.) The "foot" route seems most plausible since they all noted the presence of adult females in the foot tissues but only once in association with the gut based on greater than 4000 snail necropsies by Camp (2007) and Zimmermann (2010).

Using data acquired in the field and measurements of larval stages collected during snail necropsies, Mike concluded that adult female *D. potomaca* leave a snail host and are quickly recruited by another snail. Careful necropsies by Lauren and Mike indicated that all developmental stages of both sexes were present at any given time. He confirmed Chernin that shedding involved almost entirely adult females. Over the course of his work, Mike counted 15,532 shed females from a total of 15,641 nematodes (>99%) that he sexed. In addition to eggs in the tissues, there was also a relatively narrow assortment of larvae. In most cases, only one egg was present in gravid females; four eggs were the most seen in a single female. Three distinct nematode size classes were observed in snail tissues. In the smallest group, the gender of the parasite could not be discerned. However, in the next size class (47 males (8%) and 515 females (92%)), gender could be identified via distinct differences in tail morphology (Figure 12.1a, b). Among the largest size class (adults), there were 515 males (~7%) and 6684 females (~93%). Based on these observations, Mike concluded that the L_3 larval stage emerged from the egg and molted twice to become an adult inside the same snail in which the eggs were deposited.

Along with the results of Chernin et al. (1960), Chernin (1962), Camp (2007), and his own work (Zimmermann, 2010), Mike finally concluded that female *D. potomaca* exited the host snail and then reentered another snail. Based on these findings, he posited that eggs were shed while females were inside the host

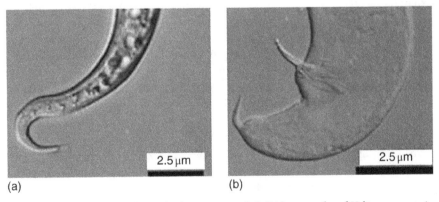

(a)   (b)

Figure 12.1   Male and female *Daubaylia potomaca*, rhabditid nematodes of *Helisoma anceps* in Mallard Lake, North Carolina. (a) The females are slightly larger than males and are easily distinguished by the hooklike posterior end of the body. (b) Males possess a gubernaculum and spicule. Zimmermann et al. (2011). Reproduced with permission of American Society of Parasitologists.

snail; the L_3 stage developed inside the eggshell and then hatched, molted twice, and became adult males and females. Although *D. potomaca* belongs in the family Cephalobidae, the suggested life cycle for the nematode is most unusual for species in this family. In other species within the family, L_1's emerge from eggs and molt four times to become adults.

In addition to the atypical life cycle of *D. potomaca*, the nematode exhibits several other unique life history traits. For example, it is not unusual for numerical aggregation in parasite infrapopulations to decline over the winter months. This phenomenon has been well documented in several other snail hosts (Fernandez and Esch, 1991). In Mallard Lake, however, the variance/mean ratios of adult *D. potomaca* in *Hel. anceps* went from being highly overdispersed in November to random by March. Moreover, the prevalence of the nematode in snail hosts also increased from 10.3% in November to 47.3% in March. Parallel with the decline in nematode infection levels is a decline in the intensity of echinostome metacercariae, which also use the snail as a host. The only reasonable explanation for the significant changes in prevalence and overdispersion of *D. potomaca* can be related to host mortality and simultaneous recruitment of the nematode. If this is correct, then it suggests that the free-living adult nematodes are active during the winter months when water temperatures are relatively low and are actually being recruited by snails that are in a state of torpor in the substrata of the lake.

During Mike's fieldwork, he encountered another symbiotic organism associated with *Hel. anceps* in Mallard Lake. The symbiont was *Chaetogaster limnaei limnaei*, a cosmopolitan oligochaete annelid that occupies the mantle cavity of freshwater snails and clams throughout the world. The functional association of the oligochaete has been depicted in several ways, for example, symbiont, commensal symbiont, parasite, predator, etc. I believe that the identity of the association depends to some extent on the objective(s) of the author(s) involved in the study. We know that the oligochaete is a gape-limited predator, which will consume a relatively wide range of microorganisms that are present in freshwater systems, including protozoans, rotifers, nematodes, and other planktonic organisms. Another important food item includes trematode miracidia that are attempting to enter snail tissues, as well as trematode cercariae that are exiting the snail. While this does not resolve the issue of the functional relationship, it did permit Mike to exploit another avenue of investigation.

The same snail species and *C. l. limnaei* were present in two systems not more than 38 km apart, that is, Charlie's Pond (Fernandez et al., 1991) and Mallard Lake (Zimmermann, 2010; Zimmermann et al., 2011). These two settings provided Mike with an extraordinary opportunity for comparing the very different parasite population dynamics and community structure affected by the oligochaete, and other species interactions, in the two snail populations. When the present study was initiated in Mallard Lake, Mike quickly learned that *C. l. limnaei* and echinostome rediae do not interact—they can't, because the annelid occupies the mantle cavity and the trematode rediae are inside the snail, usually in the hepatopancreas or gonads. However, echinostome cercariae are known to encyst as metacercariae in the mantle cavity and internally in some snail species. If *C. l. limnaei* occupies the mantle cavity of *Hel. anceps*, there is an obvious opportunity for the annelid to ingest cercariae of the echinostome, and they were observed to do so. Nonetheless, the success of this predation is mitigated because *C. l. limnaei* are gape-limited predators and echinostome cercariae are relatively large. This does not mean that the predation on echinostome cercariae by the annelid does not occur, but no more than one cercaria at a time was ever observed in *C. l. limnaei*. Mike also produced evidence that the annelid affects the prevalence of echinostomes in Mallard Lake *Hel. anceps*. His data indicated that *C. l. limnaei* directly affected echinostomes via predation on their miracidia, as well as miracidia of the other trematodes infecting *Hel. anceps*. Moreover, the evidence presented by Mike also suggests that predation by echinostome rediae in the tissues negatively affects the population biology of the nematode in *Hel. anceps* via consumption of both larvae and adults of *D. potomaca*.

The situation regarding the impact of *C. l. limnaei* on the trematodes in Charlie's Pond is different from that which occurs in Mallard Lake. Part of the difference is related to the absence of *D. potomaca* in Charlie's Pond, plus the presence of a particular species of trematode that is found in the latter body of water, but not in Mallard Lake (refer to Chapters 6 and 7). *Halipegus occidualis* was the dominant trematode in *Hel. anceps* snails from Charlie's Pond during the most years of our work. The unique aspect of the relationship between the annelid and *H. occidualis* is related to the fact that the trematode's cystophorous cercariae do not swim and that thousands are released from a single snail on a daily basis (Macy et al., 1960; Shostak and Esch, 1990); they are quite vulnerable to predation/consumption by the mantle-dwelling annelid. While generally just two to three *H. occidualis* cercariae are visible at a time inside the annelid, as many as 20 cercariae have been counted in a single *C. l. limnaei* in Charlie's Pond (Fernandez et al., 1991). Infrapopulations of *C. l. limnaei* were usually higher in snails infected by *H. occidualis*, probably due to availability of their nonmotile cercariae as an abundant food resource.

Snails possessing *C. l. limnaei* shed cercariae of other trematode species, but their impact on the population biology of the annelid was negligible. It is of interest to note, however, that cercariae of *Tylodelphys* (*Diplostomulum*) *scheuringi* were never observed (Fernandez et al., 1991) in the gut of *C. l. limnaei*. Moreover, the latter authors revealed that the number of oligochaetes present in snails infected by *T.* (*D.*) *scheuringi* was always very low, or zero, leading to the possibility that there was some sort of a special negative relationship between this trematode species and the oligochaete. Casual laboratory observations of *T.* (*D.*) *scheuringi* cercariae and the annelids indicate that whenever the cercariae made contact with *C. l. limnaei*, there was "…a dramatic withdrawal response by the oligochaete" (Fernandez et al., 1991). They noted that the reaction on the part of the annelid was identical to that which they observed for mosquitofish when externally exposed to the same cercariae, suggesting the possible release of a powerful "stinging agent" (like a venom?) from the penetration glands of the larval trematode.

As implied previously, the original idea for the master's degree research by Kyle was not to overlap with that done by Mike; in effect, the research accomplished by one was to complement the work of the other. It was, therefore, agreed that Mike was to focus on what was happening to *D. potomaca* inside the snail and Kyle was to examine the nematode's activities outside the snail. We knew from the work of Chernin et al. (1960), Chernin (1962), and Camp (2007) that adult females of *D. potomaca* were shed from snail hosts and within a few hours would be recruited by *Hel. anceps*. Efforts to determine how recruitment occurred were unsuccessful, but we assumed that the parasite penetrated the foot of the snail. Part of our reasoning for this assumption was based on the fact that *Hel. anceps* is not a burrower in the substratum, except at the beginning of winter when the snails enter a "torpor-like" state in mud at the bottom of the lake.

So, beginning in the fall of 2008, Kyle began a methodical effort to collect surface mud from various locations in the lake in order to determine their spatial distribution. The impoundment was divided into 148 transects, with each identified according to the dominant substratum type, for example, covered by leaf litter, or not. Using a random number generator, he organized his fieldwork so that, at the end of each collecting effort, he would have "24 independent samples" from each of 12 transects (Luth, 2010). These samples were brought to the lab and carefully inspected for *D. potomaca* using the Baermann funnel technique to isolate the nematodes.

We both still feel that he should have found *D. potomaca* from the substrata in the lake, but it did not happen. Kyle estimates that at least 10 gallons of mud were examined during the course of a 12-month sampling period. Considering the size of the parasite and the quantity of mud examined, we cannot see how the worm could have been overlooked. However, it is possible that the worms are not long lived as free-living organisms and that the transfer time between snails is brief. We know that under lab conditions, recently shed female nematodes are rapidly recruited by new snails and that males are very rarely shed. Nonetheless, this still represents an enigmatic gap regarding our understanding of the biology of *D. potomaca*.

After about 6 months of collecting, and with no *D. potomaca* results, Kyle and I decided that he needed to shift directions if he was to finish his thesis and graduate on time. But what to do, and how to do it? We discussed the problem and decided that he was already doing it, that is, while looking for *D. potomaca*, he had been identifying the other free-living nematodes in the lake. He was, in effect, examining the entire nematode community, both parasitic and free living. We decided that a description of the nematode community in Mallard Lake would be a good master's thesis and that, moreover, he might still encounter the elusive free-living *D. potomaca* in substrata/water interface at the bottom of the impoundment. He was very careful in establishing the baseline goal of his study because he recognized that the work could move in several directions and get out of hand as a consequence. So, he defined the objective to "...determine the influence of spatiotemporal factors on the community composition of the free-living nematofauna in this system" (Luth, 2010).

When I teach parasitology and arrive at the section dealing with nematodes, I always spend some time lecturing first about free-living species of nematodes, followed by nematode parasites. Part of my reason for using this "tack" is related to their (i) morphological diversity (while superficially they are similar in many ways, biologically they are tremendously variable—considering how many different vertebrate and invertebrate hosts they are known to infect, as well as all of the different sites of infection within these hosts), (ii) abundance (20,000–25,000 known species, with about a million remaining to be described), (iii) geographic distribution (in the deepest ocean, on the highest mountain, at the bottom of the Arctic and Antarctic waters where temperatures are always close to zero, and in the hot springs of Yellowstone where temperatures are >95°C), (iv) lifestyle versatility (as free-living species under bar mats in Bavarian pubs or on the ocean floor where nematodes represent about 90% of the life forms or parasites of both plants and animals), (v) and so on.

An apt description for the distribution of nematodes on "Mother Earth" was provided by a great nematologist of the late 19th and early 20th centuries, Nathan Cobb:

> In short, if all the matter in the universe except nematodes were swept away, our world would still be dimly recognizable, and if, as disembodied spirits, we could then investigate it, we should find its hills, vales, rivers, lakes, and oceans represented by a film of nematodes. The location of towns would be decipherable, since for every massing of human beings there would be a massing of certain nematodes. They would still stand in ghostly rows representing our streets and highways. The location of various plants and animals would still be decipherable, and, had we sufficient knowledge, in many cases even their species could be determined by an examination of their erstwhile nematode parasites. (Cobb, 1914)

The nematodes Kyle collected were classified in three ways. First, they were identified as morphotypes and, if possible, as specific taxa down to the generic level, but no further. Second, he was able to distinguish five of the eight feeding types described by Yeates et al. (1993), that is, plant feeders, bacterial feeders, predators of animals, unicellular eukaryotic feeders, and omnivores (Kyle added parasitic morphs, "...because this group was of special interest to the author" [Luth, 2010]). Then, based on buccal morphology, the morphotyped nematodes were grouped as deposit feeders/swallowers, epistrate feeders, chewers, suction feeders, or parasites.

Altogether, he collected 2349 nematodes and identified 1928 to morphotype. The discrepancy in numbers is largely based on the change in thrust of the study as described previously. Thirty genera were identified, with five representing approximately 77% of the total number found. Approximately 45% of the nematode community was identified as deposit feeders or swallowers; these morphotypes ingest and swallow their food whole (Traunspurger, 1996) and were the most prevalent and abundant in the lake. The least abundant group included parasitic morphs, all of which belonged in the order Mermithida.

There was a distinct seasonal change in prevalence and abundance within the free-living nematode community. Kyle emphasized that these changes were most likely reflections of changes in the trophic dynamics within the nematode community as a whole. Thus, for example, his observations indicated that primary consumers increased in abundance from winter to spring and predators followed the same pattern. Substratum type also was a substantive influence on spatial distribution. The most widespread nematodes were bacterivores; they were also the most prevalent feeding type on all but one substratum type in the lake. They were also the most abundant morphotype.

By far, the most dominant taxon observed in Kyle's study was *Monhystera* sp. Representatives of this bacterivore dominated nearly every month, season, and substratum type, not unlike results from several other studies (Prejs, 1977; Strayer, 1985; Wu and Liang, 1999; Eisendle, 2008). To establish an even better resolution of the community structure, Kyle devised a system based on two independent feeding classifications. One was based on the type of food consumed by each genus present in the lake. The second focused on the method of food acquisition, which, in turn, was based on buccal morphology. While using these finer scales, it became clear to Kyle (Luth, 2010) that season, trophic level, and substratum type all played significant roles in structuring the nematode community in Mallard Lake.

Both Mike and Kyle finished their master's degrees in the normal 2-year time period. Not long before completing their thesis work, I approached each of them individually, with a proposition/idea for continuing on for the PhD degrees here at Wake Forest. As described earlier (Chapter 3), I had conducted some research on the cestode *Proteocephalus ambloplitis*, which occurs in small—and largemouth—bass in North America. My interest in this particular parasite goes back nearly 45 years when I first encountered the tapeworm while working in Gull Lake, Michigan. My interest in the cestode was intensified by the work of Herman Eure (1976) in Par Pond, South Carolina, while conducting his dissertation research at the Savannah River Ecology Laboratory (SREL). Herman's results revealed that adults of the parasite were present during the winter months in the south, while I had observed them in bass during the summer months in the north. Since the cestodes in the north and south are morphologically similar, the question that emerged was, are these cryptic species, or is there an "ecological switch" in the transmission process somewhere in between the northern and southern populations? I thought the best way to resolve the question would be to subject the parasites to PCR analysis and sequencing. I approached Kyle with the idea and he jumped at it.

I wanted both Mike and Kyle to stay with me for their PhD research, so we had to come up with an idea for Mike. In Chapter 5, I described a concept that Al Bush, John Aho, Clive Kennedy, and I conceived (Esch et al., 1988), which focuses on the ecology of parasites that complete their life cycles within the confines of a single pond or lake (autogenic species) versus those that employ definitive hosts, which move from one body of water to another (allogenic species). While "concocting" this idea with Al, John, and Clive, I frequently wondered about how certain aspects of the population genetics of these two groups of parasites compared with each other. One of the ideal parasites to use in comparing these lifestyles was *P. ambloplitis* (clearly autogenic), and I already had Kyle "hooked" for this one. The allogenic species of parasite was also easy (*Echinostoma* spp.), in fact, probably the best for comparative purposes and for a number of reasons. For example, Mike would not need to collect vertebrate definitive hosts to harvest specimens for molecular analysis (which would require at lot of collecting permits in a lot of states). Instead, all he had to do was to collect snails and metacercariae encysted in the mantle cavity—either metacercariae or adults would be satisfactory for the population genetics! Kyle's genetics work on *P. ambloplitis* would give them each a target for comparative observations plus provide a specific focus for examination of population biology and genetics of a helminth parasite. So, I approached Mike with the proposition and he quickly agreed!

Both of the guys have been working on their dissertation for 3+ years as of today's writing. Mike is doing his PCR and sequencing work at the present time. Kyle is about to start. Why, you might ask, has it taken so long for them to get to this point? The answer is easy. As I explained earlier, they both decided to expand their collecting effort (which took 3 years) with funding support from the Department of Biology's Grady Britt Fund.

Kyle and Mike resolved that before they pursued the PCR/sequencing part of their study, they would collect in several states east of the Mississippi River in 2012, but they also decided that was not enough. So, they expanded their collecting efforts in 2013 all the way to the Rocky Mountain states of Wyoming, Colorado, and New Mexico, south to include all of the Gulf Coast states except Florida, and to the north, including Michigan, Wisconsin, Minnesota, and North Dakota. During the past year, Kyle persuaded Joanna Reinhold and Amanda Rosenfeld, two of my undergraduate students in the parasitology course that I taught in the spring of 2013, to come on board as full partners. They went through the intestines of all the fish Kyle necropsied, identified the parasites present, and counted them as well. They both will be made coauthors on at least two of the papers that will emerge based on Kyle's collecting efforts.

I am confident that their collections represent two of the largest sample sizes and widest geographic distributions of any parasite study ever attempted in North America. As of this moment in writing the book, Kyle's 7-month-long necropsy effort is just short of completion. The data analysis will be underway shortly. Mike will be observing data from his first PCR/sequencing efforts very soon. I'll be anxious to see their results—so will they!

References

Camp, L.E. 2007. Infection dynamics and life cycle of Daubaylia potomaca (Nematoda: Rhabdita) in Helisoma anceps. Master's thesis. Wake Forest University, Winston-Salem, North Carolina, 61 p.

Chernin, E. 1962. The unusual life history of *Daubaylia potomaca* (Nematoda: Cephalobidae) in *Australorbis glabratus* and in certain other fresh-water snails. *Parasitology* **52**: 459–481.

Chernin, E., E.H. Michelson, and D.L. Augustine. 1960. *Daubaylia potomaca*, a nematode parasite of *Helisoma trivolvis*, transmissible to *Australorbis glabratus*. *Journal of Parasitology* **46**: 599–607.

Chitwood, B.G., and M.B. Chitwood. 1934. *Daubaylia potomaca*, n.sp., a nematode of snails, with a note on other nemas associated with molluscs. *Proceedings of the Helminthological Society of Washington* **1**: 8–9.

Cobb, N. 1914. Nematodes and their relationships. United States Department of Agriculture, Beltsville, MD, p. 457–490.

Eisendle, U. 2008. Spatiotemporal distribution of free-living nematodes in glacial fed stream reaches (Hohe Tauern, Eastern Alps, Austria). *Arctic, Antarctic and Alpine Research* **40**: 470–480.

Esch, G.W., C.R. Kennedy, A.O. Bush, and J.M. Aho. 1988. Patterns of helminth communities in freshwater fish in Great Britain: Alternative strategies for colonization. *Parasitology* **96**: 519–532.

Eure, H.E. 1976. Seasonal abundance of *Neoechinorhynchus cylindratus* taken from large-mouth bass (*Micropterus salmoides*) in a heated reservoir. *Parasitology* **73**: 355–370.

Fernandez, J.C., and G.W. Esch. 1991. Guild structure of larval trematodes in the snail *Helisoma anceps*: Patterns and processes at the individual host level. *Journal of Parasitology* **77**: 528–539.

Fernandez, J.C., T.M. Goater, and G.W. Esch. 1991. The population dynamics of *Chaetogaster limnaei limnaei* (Oligochaeta) as affected by a trematode parasite in *Helisoma anceps* (Gastropoda). *American Midland Naturalist* **125**: 195–205.

Luth, K.E. 2010. Influence of spatiotemporal variables in structuring the nematode community of a fresh water system. Master's thesis. Wake Forest University, Winston-Salem, North Carolina, 175 p.

Macy, R.W., W.A. Cook, and W.R. DeMott. 1960. Studies on the life history of *Halipegus occidualis* Stafford, 1905 (Trematoda: Hemiuridae). *Northwest Science* **34**: 1–17.

Prejs, K. 1977. The littoral and profundal benthic nematodes of lakes with different trophy. *Ekologica Polanica* **25**: 21–30.

Shostak, A.W., and G.W. Esch. 1990. Photocycle-dependent emergence by cercariae of *Halpegus occidualis* from *Helisoma anceps*, with special reference to cercarial emergence patterns as adaptations for transmission. *Journal of Parasitology* **76**: 790–795.

Strayer, D. 1985. The benthic micrometazoans of Mirror Lake, New Hampshire. *Archives of Microbiology* **72**(Suppl.): 287–426.

Traunspurger, W. 1996. Distribution of benthic nematodes in the littoral zone of an oligotrophic lake (Konigssee National Park Berchtesgaden, FRG). *Archives of Hydrobiology* **135**: 393–412.

Wu, J., and Y. Liang. 1999. A comparative study of benthic nematodes in two Chinese lakes with contrasting sources of primary production. *Hydrobiologia* **411**: 31–37.

Yeates, G.W., T. Bongers, R.G.M. Goede, D.W. Freckman, and M. Georgieva. 1993. Feeding habits in soil nematode families and genera—an outline for soil ecologists. *Journal of Nematologists* **25**: 315–331.

Zimmermann, M.R. 2010. Population biology of Daubaylia potomaca (Nematoda: Rhabditida) in Mallard Lake, North Carolina. Master's thesis. Wake Forest University, Winston-Salem, North Carolina, 100 p.

Zimmermann, M.R., K.E. Luth, and G.W. Esch. 2011.Complex interactions among a nematode parasite (*Daubaylia potomaca*) and a commensalistic annelid (*Chaetogaster limnaei limnaei*), and trematodes in a snail host. *Journal of Parasitology* **97**: 788–792.

13 The Catastrophic Collapse of the Larval Trematode Component Community in Charlie's Pond (North Carolina)

> *Everything is sweetened by risk.*
> *Of Death and the Fear of Flying*, Alexander Smith (1830–1867)

Writing a book takes a relatively long period of time—this one was no different. The two chapters (6 and 7) dealing with Charlie's Pond were written more than a year ago. In 2012, three undergraduate students (Collin Russell, Tina Casson, and Courtney Sump) expressed an interest in doing some research in my laboratory. So, I assigned them the task of collecting snails from Charlie's Pond and identifying cercariae that were shed after isolating the snails in the lab. Three graduate students (Kyle Luth, Mike Zimmermann, and Nick Negovetich) worked directly with the undergraduates or assisted their research in other ways. Based on these efforts, all seven of us contributed to a recent paper published in the *Journal of Parasitology* (2015, **101**: 116–120). (The author sequence for the paper is Russell, Casson, Sump,

Ecological Parasitology: Reflections on 50 Years of Research in Aquatic Ecosystems,
First Edition. Gerald W. Esch.

Luth, Zimmermann, Negovetich, and Esch.) Because I believe the results of the final study are significant to our long-term work in Charlie's Pond, I decided to contact Dr. Michael Sukhdeo, editor of the *Journal of Parasitology*, and ask for his permission to include the content of that paper as part of this book. He graciously agreed to my use of the critical comment paper, almost verbatim, and without the usual quotation marks. We enthusiastically thank him for his permission. There is some repetition here for information present in Chapters 6 and 7. However, to maintain an accurate story line, a decision was made to reiterate some of it here.

Only a few long-term studies exist in the parasitological literature. Kennedy et al. (2001) demonstrated over a 31-year period that population changes of the plerocercoid stage of *Ligula intestinalis* are intimately tied to the community dynamics involving roach (*Rutilus rutilus*), rudd (*Scardinius erythrophthalmus*), and great crested grebes (*Podiceps cristatus*). This work was performed in the Slapton Ley (Devon, England), an established lake and ecosystem that had experienced eutrophication prior to, and during, the onset of their study. The influence of eutrophication on parasite communities was also examined in a 20-year study in Gull Lake, Michigan (Marcogliese et al., 1990). There, anthropogenic eutrophication altered host distribution within the lake, thereby disrupting the natural transmission dynamics of *Crepidostomum cooperi* (Allocreadiidae). Janovy et al. (1997) studied the effect of natural disturbances on parasite communities over a 14-year period in the South Platte River in Nebraska. Streamflow had the greatest influence on the parasite community (prevalence and abundance) of *Fundulus zebrinus* through its effect on parasite transmission. While these three long-term studies documented changing parasite communities, all of the work was performed in well-established ecosystems. Thus, while fluctuations in parasite communities most accurately represent the resilience of late-stage successional stages to natural and anthropogenic disturbances, the parasite community always appears to return to predisturbance levels.

To our knowledge, long-term studies (lasting 31 years) involving component trematode communities in molluscan hosts do not exist. In 1983–1984, Amy Crews began collecting *Helisoma (Hel.) anceps* from a small, spring-fed body of water (Charlie's Pond), which drains into Belews Lake, a large cooling reservoir owned by Duke Energy in the southwest corner of Stokes County, North Carolina (Crews and Esch, 1986). The pond was formed in the early 1970s following construction of an access road to a coal-fired power plant. While a small, spring-fed creek likely existed prior to impoundment, the pond only became established after construction of the access road. As such, the newly formed pond began its transformation toward an established lentic habitat. Early hosts and parasites were those that were previously present in the lotic system or colonized after impoundment.

The first collections at Charlie's Pond were in 1983 and continued during 1984. *Helisoma anceps* snails were collected from the pond, brought to the laboratory, separated into small plastic jars containing aged tap water, and checked for the release of cercariae. During the first two years, eight species of trematodes were identified. Seven of the species present exhibited prevalence of less than 4%, reasonably representative percentages for digenetic trematodes in molluscan hosts in freshwater systems. However, a hemiurid, *Halipegus* (=*Hal.*) *occidualis*, was atypical in several respects. The prevalence of this parasite in *Hel. anceps* was consistently approximately 60% during the first few years, except for brief periods of time, always in late June and early July when the snail population was turning over; there were even a couple of brief periods when prevalence reached 100%. All snails with patent infections of *Hal. occidualis* were either castrated or on the way to becoming castrated (Crews and Esch, 1986). However, it was subsequently found (Goater et al., 1989) that approximately 25% of overwintering *Hel. anceps* infected with *Hal. occidualis* lost their infections and that many of these snails resumed egg production in the following spring.

The prevalence of *Hal. occidualis* in Charlie's Pond was relatively high over the years, although there has been a decline since the 1984–1986 time period. Prevalence was 50% in 1989–1990 (Fernandez and Esch, 1991b) and slightly higher (57%) in 1991 (Williams and Esch, 1991). In more recent years, *Hal. occidualis* prevalences were 18, 17, and 15% for 2002, 2005, and 2006, respectively (Negovetich and Esch, 2007). In the same paper, they noted a decline in overall trematode infection percentages in snails from 31.4% in 1984 to 8.7% in 2006. *Halalipegus occidualis* was present in 2012, but with a prevalence of just 4.8%. The data clearly show that prevalence of the dominant trematode in *Hel. anceps* declined significantly since the first survey by Crews and Esch (1986). Indeed, collections made in 2013 and 2014 indicated local extinction of *Hal. occidualis* in the pond.

From 1984 through 2006, 18,636 *Hel. anceps* snails were collected from Charlie's Pond as part of either a master's or PhD research project. Throughout this time period, 18 trematode species were identified; *Hal. occidualis* was present during each year (through 2006) in which snails were collected, albeit with declining frequency. *Megalodiscus temperatus* was present in each year except the first. Both of these species are autogenic, with each using green frogs (*Rana clamitans*) as the definitive host. The next most frequently observed species was an echinostome (allogenic), present in each year except 2001. Some allogenic species were sporadically recovered, for example, *Petasiger nitidus*, *Clinostomum* sp., *Uvulifer ambloplitis*, etc., while others (*Echinostoma* sp., *Tylodelphys* (*Diplostomulum*) *scheuringi*) were consistently present (Negovetich and Esch, 2007). Overall, *Hel. anceps* was the focus of most of the research at Charlie's Pond, but other snail species, and their associated parasites, also existed in the pond and were also studied. Specifically, *Physa gyrina* currently represents the second most abundant snail species in the pond.

Physa gyrina colonized the pond sometime during the first two years of research. Goater (1989) observed the occurrence of *Halipegus* (*Hal.*) *eccentricus* in green frogs in the pond serendipitously during 1986 and confirmed its identification with the discovery of *Hal. eccentricus*-infected *P. gyrina*. During its first summer, *Hal. eccentricus* prevalence in snails was 49% (Goater, 1989). *Halipegus eccentricus* prevalence peaked at approximately 50% for Snyder (1992) in 1991–1992 and approximately 30% for Sapp (1993) in 1992–1993. The prevalence of *Hal. eccentricus* in green frogs was approximately 55% in 1993 (Wetzel, 1995). However, in 1994 and 1995, just two green frogs from 180 captures were infected with *Hal. eccentricus* (D.A. Zelmer, personal communication), representing a relatively rapid decline in prevalence. The parasite was not observed in a collection of approximately 2400 *P. gyrina* from Charlie's Pond during a 9-month period in 1996 (Schotthoefer, 1998) nor was it observed during 2012–2014 in a collection of 983 *P. gyrina*.

The Snyder (1992) and Sapp (1993) reports are the most comprehensive for the component trematode community in *P. gyrina* from Charlie's Pond. Snyder obtained data between April 1991 and March 1992 from 1181 snails and Sapp between April 1992 and March 1993 for 1231 snails. Based on the data generated in these two studies, the component communities in *P. gyrina* were both rather depauperate, consisting of just six species for Snyder and seven for Sapp. In each study, the same four species were autogenic in *P. gyrina* (*Hal. eccentricus, Haematoloechus complexus, Glypthelmins quieta,* and *M. temperatus*); all four of these trematodes employ green frogs as the definitive host. The additional fluke in Sapp's work was an allogenic strigeid.

Beginning in 2011 and continuing through August of 2014, collections of both *Hel. anceps* and *P. gyrina* were made in Charlie's Pond. In 2013, only two species of trematode were observed in *Hel. anceps*, that is, *Haematoloechus (Hae.) longiplexus* and *U. ambloplitis*. The intramolluscan infracommunities of *P. gyrina* included one species of echinostome (probably *Echinoparyphium* sp.); two specimens of *P. gyrina* were infected with an unknown strigeid. The sample size for 2013 was small (275 *Hel. anceps* and 381 *P. gyrina*), so a repeat sample of snails was made in August 2014. Only two trematodes, *Hae. longiplexus* and an echinostome, were recovered from 839 *Hel. anceps*. Four trematode species were recovered from 267 *P. gyrina*, that is, an echinostome (7.9%), *Plagiorchis* sp. (4.9%), *G. quieta* (1.5%), and *P. minimum* (7.4%). The low species richness for both 2013 and 2014 strongly suggest that the component community of *Hel. anceps* has changed since Crews first collected in the pond.

Of the 18 species that have been recorded from *Hel. anceps* since 1984, seven species were present in at least half of the years in which studies were conducted. Four of the seven were autogenic, that is, *Hal. occidualis, Hae. longiplexus, M. temperatus,* and *P. parva*. The first three of these species use green frogs as definitive hosts, while the latter is typically found in salamanders. Among the 11 allogenic species, we would expect to see at least some temporal variability within the pond. Three of them, including *T. (D.) scheuringi,* an unknown spirorchiid, and *Echinostoma* sp., were present in more than half of the sampling years. We cannot be completely certain, but it is our belief that the majority of the allogenic species were brought to the pond by fish-eating birds. Notably, trematode species richness in *Hel. anceps* appears to have declined in recent years. The 2006 collecting season was the only year when the 95% CI included the richness value of five, thus suggesting that trematode diversity has declined compared to when research first began at the pond. Furthermore, only two trematode species were recovered from *Hel. anceps* collected between April and November in 2013 ($n = 275$), and two species were identified from 839 snails collected in August 2014. None of the 95% CI extended below four for sample size of 1006 (Figure 13.1) or 839 (sample size of 2014), thus suggesting that the decline in species richness is not due to a smaller sample size. Variation in species richness will exist between the years because of the presence/absence of allogenic species, but the loss of an autogenic species, that is, *Hal. occidualis*, will cause a decrease in the mean species richness. In Charlie's Pond, the dominant autogenic species utilized frogs as definitive hosts, which is not surprising since these amphibians are permanent residents in the pond.

Goater (1989) began research on the green frog–parasite relationship in the pond in 1986; over a 3-year period, he collected 199 frogs (an average of ~66 per year). None of these frogs was killed; all were, however, toe clipped for identification at the time of recapture. We have not been involved with frog sampling since the work of Wetzel (1995) and Zelmer (1998). In 1992, Wetzel captured 80 frogs; the number captured declined to 45 in 1993 and 38 in 1994. Zelmer collected 44, 57,

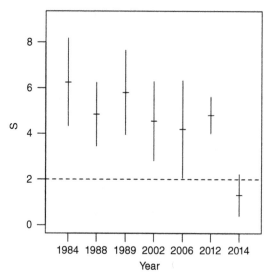

Figure 13.1 Rarefied species richness. Mean species richness and 95% confidence intervals were calculated using 2000 replicates (random samples without replacement) for a sample size of 275 (the sample size for 2013). The dashed line marks the species richness for the 2013 sample. Raw data were used for 1984, 2002, 2006, 2012, and 2014. Data for 1988 were obtained from Williams and Esch (1991). Data for 1989 were obtained from Fernandez and Esch (1991b). Russell et al. (2015). Reproduced with permission of the American Society of Parasitologists.

and 45 from 1994 through 1997. Excluding the 80 frogs captured in 1992, the overall average per year was approximately 46.

The pond has remained physically unchanged over the period from 1984 through 2014. How then do we account for the relatively steady decline in the trematode diversity in *Hel. anceps* (and, probably, in *P. gyrina*)? The gradual reduction in parasite diversity suggests that a fundamental change of some sort has occurred within the pond. We are unaware of any toxic agents that might have been added accidentally.

We have no reason to believe that the pond, as a habitat, has undergone a significant change, except for one feature. *Typha latifolia* and *Juncus effusus* (emergent vegetation) stands have come and gone several times since we began sampling in 1984. Several studies (Crews and Esch, 1986; Fernandez and Esch, 1991a; Williams and Esch, 1991; Sapp and Esch, 1994) observed heterogeneous spatial distributions of hemiurid parasites in both *P. gyrina* and *Hel. anceps*. However, Zelmer et al. (1999) clearly established that the patchy spatial distribution of *T. latifolia* and *J. effusus* was strongly correlated with infection foci of *Hal. occidualis* in the pond. The microhabitats identified in these stands of emergent vegetation were apparently conducive to transmission dynamics involving each step in this parasite's life cycle. This is not to say that the parasites were not transmitted in other locations throughout the pond, but these four were the most active sites.

Photographs (Figure 13.2a and b) taken at the pond in 1988 illustrate the presence of emergent vegetation. The photos in Figure 13.2c and d were shot in the summer of 2014 to show the absence of emergent vegetation at the same two sites (15 and 24 in Figure 2b of Zelmer et al., 1999). Regrettably, we did not keep records for the presence/absence of emergent vegetation over the years, so we cannot propose any more than a reasonable hypothesis for the local extinction of at least one of the two hemiurid species (*Hal. occidualis*) due to the loss of emergent vegetation.

In the early years of our work in the pond, there were four sites with emergent vegetation (Figure 2b in Zelmer et al., 1999). Between 1988 and 2006, when Negovetich and Esch (2007) were conducting research, at least 75% of this vegetation had been lost. Extensive drought lasting through 2002 reduced water levels in the pond may have affected the maintenance of emergent vegetation. Furthermore, the water level in the pond declined about 30 cm in April 2006 when debris blocking a pipe that drained excess water from the pond was cleared. These drying events, in conjunction with the accumulation of leaf litter, may be responsible for the gradual decrease in the amount of emergent vegetation and the subsequent effect on the transmission dynamics of the digenetic trematodes at Charlie's Pond.

Zelmer et al. (1999) clearly established the existence of four *Hel. anceps* transmission "hotspots" in the pond, all associated with emergent vegetation. We believe that these sites supported increased densities of *Hel. anceps* because of the increased size of aufwuchs (periphyton) communities on the submerged parts of the vegetation; these aufwuchs communities likely encouraged an increase in ostracod population densities as well. The emergent vegetation offered excellent sites for eclosion by

Figure 13.2   (a) East bank of Charlie's Pond showing emergent vegetation in 1988. (This location matches site 24 in Zelmer et al. (1999).) (b) West cove of Charlie's Pond showing vegetation in 1988. (This location matches site 15 in Zelmer et al. (1999).) (c) East bank of Charlie's Pond without vegetation in 2014. (d) Northwest cove of Charlie's Pond without vegetation in 2014. Russell et al. (2015). Reproduced with permission of the American Society of Parasitologists.

dragonflies during their life cycles. The green frogs, as ambush predators, were provided exceptional opportunities for preying upon emerging dragonfly naiads, as well as protection from foraging predators. In other words, everything necessary for completing the life cycle of *Hal. occidualis* was not only present in these four sites; it was concentrated there as well. In 2013 and 2014 (Figure 13.2c and d), emergent vegetation was not present.

While we suggest a possible link between the loss of the vegetation and the decline and disappearance of *Hal. occidualis*, there are other possible explanations as well. For example, frog population size could have dropped to such low numbers as to take the parasite infrapopulation sizes below a sustainable number. We are certain the frog population in Charlie's Pond declined over a 10-year period (Zelmer et al., 1999) and that amphibian populations have declined in many parts of the world. However, recent cursory observations by us, and by others, indicate that green frogs are still in Charlie's Pond. While we have never checked the frog population for chytrid fungi, we have no evidence indicating that neither it nor

Ribeiroia spp. infections have occurred in the pond, both of which are thought to contribute to localized problems for frog populations in many areas. Moreover, all of the trematodes in *Hel. anceps*, save three species present during 2013 and 2014, have disappeared. Of these three species, only one is autogenic, *Hae. longiplexus*. We know from 31 years of experience in the pond that allogenic species are ephemeral for the most part, but not most of the autogenic trematodes. The latter, once established, should remain as part of a consistent autogenic community, although *Hal. eccentricus* appeared, abundantly, in 1986 (Goater, 1989) and became locally extinct by 1996 (Schotthoefer, 1998).

One other idea should also be considered. Snails in general exhibit a high degree of plasticity in various life history traits. Infected snails are most likely to be larger than uninfected snails in this system and many others. For *Hel. anceps* in Charlie's Pond, the correlation between prevalence and host size is most likely due to larger snails being older and exposed to infective stages over a longer period of time (Negovetich and Esch, 2008a). Since 1984, average snail size has gradually decreased. One might argue that smaller snails are less likely to become infected because they are younger than larger snails, as supported by size differences between infected and uninfected snails in some freshwater and marine snail–trematode systems (reviewed in Sorensen and Minchella (2001)). We contend here that snail size is not a primary factor causing a decrease in trematode diversity because the smaller snails in 2006, for example, are of a similar age as those collected in the same month in previous years and have thus been exposed to trematode transmission stages for the same time period.

Size of freshwater snails is determined, in part, by nutrient quality and abundance. A thorough study by Keas and Esch (1996) suggests that snails in Charlie's Pond are nutrient deprived. This conclusion was partially based on comparing growth rates of snails raised on low-nutrient diets to estimates derived from mark–recapture studies in the field. Negovetich and Esch (2008a) also calculated growth rates from a mark–recapture study. Compared to values reported by Fernandez and Esch (1991c), snails in the spring–summer–fall of 2005–2006 grew more slowly than those from 1989 for all months except August. Negovetich and Esch (2008a) did not separate snails by maturity prior to calculating specific growth rates (snails as small as 6 mm reproduced in 2006). Both studies demonstrate that growth rates are dependent on initial size. However, snails in 2005–2006 were smaller than those in 1989. As such, it is not surprising that the 2005–2006 growth rate is greater than that in 1989 because the population in 2005–2006 consisted of a larger frequency of smaller individuals with faster growth rates. We propose that the declining growth rates are due, in part, to a declining food source present in the aufwuchs communities, which require this substratum to grow. Because emergent vegetation was lost, the amount of aufwuchs available for snails declined and, therefore, growth rates of the snail hosts also declined.

Differences in growth rates cause similarly aged snails to differ in their overall size. Another line of evidence supports our claim that the current population of

Hel. anceps is not younger than in years past, that is, size histograms. Cohort turnover of *Hel. anceps* consistently occurs during late June in the pond. If snails are not living as long, then cohort turnover should occur earlier in the year. A decrease in the age of the snails would gradually shift turnover to April and May, yet this has not occurred. Moreover, death at an earlier age has the potential to introduce a second cohort turnover later in the year, such as in August or September. The influence on the size histograms would be the appearance of bimodal distributions multiple times during the year. Gradually, the monthly size histograms should lose all appearance of a cohort turnover because individuals would be continuously replaced throughout the year. In contrast, *P. gyrina* exhibits such a size histogram. During the summer, *P. gyrina* lives for only 3–4 months (Snyder, 1992; Snyder and Esch, 1993). Snails continuously reproduce and replace the deceased individuals. As such, monthly size histograms do not reveal any indication of an annual cohort turnover.

Discussion to this point has revolved around the disruption of transmission as an explanation for the significant decline of prevalence of *Hal. occidualis* and overall loss of trematode species richness. At a single point in time, the probability of transmission depends on the number of infective stages in the habitat. The probability of becoming infected is the cumulative function describing the probability of transmission at infinitesimally small time intervals from birth to a time point of interest. Prevalence utilizes the probability of infection and, as such, depends on the number of infective stages in the environment and the time spent in the habitat, that is, age of the host. While this is a somewhat simplistic view, because the statement assumes that infective stages are randomly distributed in the habitat, we can refine this statement by defining "habitat" as the area most likely to contain all hosts in the life cycle of the parasite. In so doing, we propose the following for reasons given above. Life expectancy of *Hel. anceps* has not changed, so age of the snails is unlikely to have disrupted transmission or significantly influenced the prevalence recorded from surveys in the pond. Emergent vegetation, however, has declined, even disappeared at various times, and this has immediate impacts by reducing the probability of overlap of the frog and snail host. Additionally, the loss of plant or leaf substrata affects the availability of aufwuchs, which has been shown to affect both growth rates of the snail and cercariae production (Keas and Esch, 1996), further affecting snail population size and disrupting the trematode life cycles. We propose that emergent vegetation is central to the establishment and persistence of digenetic trematodes in aquatic systems.

We cannot, of course, be certain that our evidence supports our hypothesis regarding the impact of change in emergent vegetation within the pond. In part, this is because we do not have a control system with which to compare our data over time. Regrettably, there are also gaps in our data collections. Moreover, we have no way of comparing our long-term observations with other long-term studies because none exist. We hope that in 5 or 10 years, however, additional collections of snails from Charlie's Pond would be made so that data can be compared. We feel that we have created a solid "footprint" that should be followed in the future.

Finally, we want to recognize the hard work in Charlie's Pond performed by a large cadre of Wake Forest graduate students over the years, beginning in 1984. Their research is obviously responsible for the present paper.

References

Crews, A.E., and G.W. Esch. 1986. Studies on the population biology of *Halipegus occidualis* (Hemiuridae) in freshwater snail host *Helisoma anceps* (Pulmonata). *Journal of Parasitology* 72: 646–651.

Fernandez, J.C., and G.W. Esch. 1991a. Guild structure of larval trematodes in the snail *Helisoma anceps*: Patterns and processes at the individual host level. *Journal of Parasitology* 77: 528–539.

Fernandez, J.C., and G.W. Esch. 1991b. Component community structure of trematodes in the pulmonate snail, *Helisoma anceps*. *Journal of Parasitology* 77: 540–550.

Fernandez, J.C., and G.W. Esch. 1991c. Effect of parasitism on the growth rate of the pulmonate snail *Helisoma anceps*. *Journal of Parasitology* 77: 937–944.

Goater, T.M. 1989. The morphology, life history, ecology, and genetics of Halipegus occidualis (Trematoda: Hemiuridae) in molluscan and amphibian hosts. Ph.D. dissertation. Wake Forest University, Winston-Salem, North Carolina, 151 p.

Goater, T.M., A.W. Shostak, J.A. Williams, and G.W. Esch. 1989. A mark-release-recapture study of trematode parasitism in over-wintered *Helisoma anceps* (Pulmonata), with special reference to *Halipegus occidualis* (Hemiuridae). *Journal of Parasitology* 75: 249–257.

Janovy, Jr., J., S.D. Snyder, and R.E. Clopton. 1997. Evolutionary constraints on population structure: The parasites of *Fundulus zebrinus* (Pisces: Cyprinodontidae) in the South Platte River of Nebraska. *Journal of Parasitology* 83: 584–592.

Keas, B.W., and G.W. Esch. 1996. The role of diet amd reproductive maturity on the growth *Helisoma anceps* (Pulmonata) infected by *Halipegus occidualis* (Trematoda). *Journal of Parasitology* 83: 96–104.

Kennedy, C.R., P.C. Shears, and J.A. Shears. 2001. Long-term dynamics of *Ligula intestinalis* and roach *Rutilus rutilus*: A study of three epizootic cycles over thirty-one years. *Parasitology* 123: 257–269.

Marcogliese, D.J., T.M. Goater, and G.W. Esch. 1990. *Crepidostomum cooperi* (Allocreadiidae) in the burrowing mayfly, *Hexagenia limbata* (Ephemeroptera) related to trophic status of a lake. *American Midland Naturalist* 124: 309–317.

Negovetich, N.J., and G.W. Esch. 2007. Long-term analysis of Charlie's Pond: Fecundity and trematode communities of *Helisoma anceps*. *Journal of Parasitology* 93: 1311–1318.

Negovetich, N.J., and G.W. Esch. 2008a. Life history cost of trematode infection in *Helisoma anceps* using mark-recapture in Charlie's Pond. *Journal of Parasitology* 94: 314–325.

Negovetich, N.J., and G.W. Esch. 2008b. Quantitative estimation of the cost of parasitic castration in a *Helisoma anceps* population using a matrix population model. *Journal of Parasitology* 94: 1022–1030.

Russell, C., T. Casson, C. Sump, K. Luth, M. Zimmermann, N. Negovetich, and G. Esch. 2015. The catastrophic collapse of the larval trematode component community in Charlie's Pond (North Carolina). *Journal of Parasitology* **101**: 116–120.

Sapp, K.K. 1993. Effects of microhabitat on the transmission of larval trematodes in Helisoma anceps and Physa gyrina, and the impact of infracommunity manipulation on community structure. M.S. thesis. Wake Forest University, Winston-Salem, North Carolina, 60 p.

Sapp, K.K., and G.W. Esch. 1994. The effects of spatial and temporal heterogeneity as structuring forces for parasite communities in *Helisoma anceps* and *Physa gyrina*. *American Midland Naturalist* **132**: 91–103.

Schotthoefer, A.M. 1998. Spatial variation in trematode infections and fluctuations in component community composition over the long-term in the snails, Physa gyrina and Helisoma anceps. M.S. thesis. Wake Forest University, Winston-Salem, North Carolina, 72 p.

Snyder, S.D. 1992. Trematode community structure in the pulmonate snail Physa gyrina and the effect of parasitism on the fecundity of the snail host. M.S. thesis. Wake Forest University, Winston-Salem, North Carolina, 77 p.

Snyder, S.D., and G.W. Esch. 1993. Trematode community structure in the pulmonate snails *Physa gyrina*. *Journal of Parasitology* **79**: 205–215.

Sorensen, R.E., and D.J. Minchella. 2001. Snail-trematode life history interactions: Past trends and future directions. *Parasitology* **123**: S3–S18.

Wetzel, E.J. 1995. Seasonal recruitment and infection dynamics of Halipegus occidualis and Halipegus eccentricus (Digenea: Hemiuridae) in their arthropod and amphibian hosts. Ph.D. dissertation. Wake Forest University, Winston-Salem, North Carolina, 120 p.

Williams, J.A., and G.W. Esch. 1991. Infra- and component community dynamics in the pulmonate snail, *Helisoma anceps*, with special emphasis on the hemiurid trematode, *Halipegus occidualis*. *Journal of Parasitology* **77**: 247–253.

Zelmer, D.A. 1998. Life history and transmission dynamics of Halipegus occidualis (Digenea: Hemiuridae). Ph.D. dissertation. Wake Forest University, Winston-Salem, North Carolina, 164 p.

Zelmer, D.A., E.J. Wetzel, and G.W. Esch. 1999. The role of habitat in structuring *Halipegus occidualis* metapopulations in the green frog. *Journal of Parasitology* **85**: 19–24.

14 An Epilogue: What's Involved with Graduate School?

In this world, there are only two tragedies. One is not getting what one wants, and the other is getting it.

Lady Windermere's Fan, George Bernard Shaw (1856–1950)

All of us must make our way through a somewhat rigid academic "minefield" in order to secure a graduate degree. Having been the dean of our graduate school here for 6 years, I learned a lot about the "ins and outs" of how students think about their experiences while making their way. Since this book is aimed primarily at graduate students and younger faculty, Kyle Luth, my present and final graduate student, suggested that it would be a good idea to alert new students about what to expect when they enter a graduate program. Accordingly, the following comments are directed at those students who are just beginning or who are thinking about applying to graduate school. My intention here is to offer insight on what to expect in graduate school and how to successfully meet challenges of the "minefield."

In some places of the present chapter, I will refer to coursework hours required for a degree, or the number of faculty members serving on an advisory committee, or the time necessary to complete an exam, etc. I should make clear that these specifics apply to requirements we have here at Wake Forest, which means there will be some differences from one institution to another.

Ecological Parasitology: Reflections on 50 Years of Research in Aquatic Ecosystems, First Edition. Gerald W. Esch.

What to look for if you are being recruited

Recruiting a student would seem to be one of the easiest things to do since all of us who have made it into graduate school have experienced the process of recruitment at one time or another. I would bet, however, that there are many differences between the ways in which most of us have been pursued. However, I would also wager that many of us might actually recruit ourselves. If we are lucky enough to know what we want in terms of what we want to study, the going is easy.

My entry into graduate school was probably, however, a little different than most. At first, I wanted to study anatomy and even had a teaching assistantship in "my back pocket." I was on my way to the University of Kansas School of Medicine, at least until I was told that I had to first take a course in human gross anatomy. This woke me up and I quickly gave up the idea. At the same time I abandoned anatomy, I was fortunate to be taking a parasitology course from one of the most dynamic parasitologists of his era, Dr. Robert (Doc) Stabler. Because of him, I had developed a huge interest in parasites. He was a fantastic help in understanding what I wanted, and needed, to do, even though I was very late in applying. Applications like this should be started in early fall.

It was getting late by this time in the spring of my senior year at Colorado College, and, in my case, I had to get moving. Another serious problem was that I had no real idea about the kind of parasites I wanted to study, where I wanted to study them, or with whom I wanted to study. Fortunately, Doc gave me a membership roster for the American Society of Parasitologists and told me to pick some names. He immediately sent letters of recommendation to my selections, and within 5 days, he received a telephone call from Professor J. Teague Self at the University of Oklahoma saying that he was willing to take me into his lab. I was very fortunate, but I did not know how lucky at the time.

Over my years at Wake Forest, I have had 43 graduate students successfully pass through my lab. From where did they come and why did they come here? My very first student, John Trainer, was waiting for me when I arrived at Wake Forest. He had just graduated from Muhlenberg College, a very fine liberal arts school in Pennsylvania. I do not recall discussing it with John, but I suspect he chose Wake Forest because it was a relatively small liberal arts setting, which resembled Muhlenberg in many ways. Several years later, when I became dean of our graduate school, I conducted a poll for all of our graduate students. One of the questions I asked was, why did you come to Wake Forest? The majority of them chose us because they wanted a small school with a strong academic reputation, plus the opportunity of working directly with a professor. Their reason for coming here was really not that much different than most of our undergraduates. Wake Forest is the smallest NCAA Division I university in the United States. We are financially well endowed, our faculty to student ratio is small, our facilities are excellent, our professors are committed to both teaching and research, and many of our office doors are always open when, and if, our students want to see us. In fact, I do not have office hours. I have told my students that I will "drop" whatever I am doing and give them as much time as they want, whenever they want it.

Early in my experience at Wake Forest, our biology department would accept prospective students at the master's level if their grades were good, their Graduate Record Exam scores were high, and they had strong letters of recommendation. In addition to these criteria, a major professor also would have to personally approve the Graduate Committee's recommendation for admission at the PhD level for a given student. Today, the same academic qualifications are used. Following a current trend of many universities, however, all of our prospective graduate students are now brought to campus and interviewed directly.

One of the most important criteria for selecting a graduate school is the knowledge of the faculty member with whom a student is going to study. I personally think this factor is the most important requirement of all. A good word that describes this prerequisite is compatibility. If a student cannot "connect" with their major professor, they will find themselves in great trouble at some point during their graduate school training.

Planning your program

Coursework

A master's student

Once a student is accepted and enters a program, there are certain requirements that must be met in order to graduate with an MA or MS degree (by the way, an MA is traditionally the higher of the two degrees). A committee that includes the advisor plus three other faculty members is appointed early. This committee is responsible for approving the student's coursework and a thesis topic. Generally, the committee members should also be able to provide direct assistance in the student's area of research. Our MS students are required to take 24 semester hours of regular coursework, and they automatically receive 6 hours for their thesis research, a total of 30 credit hours. Twelve of the twenty-four regular course hours are at the graduate level, and these courses are restricted to graduate students. To complete their graduation requirement, a master's student at Wake Forest also must present a departmental seminar at the end of their fourth semester and must then defend their thesis via their own advisory committee.

A PhD student

These students are not constrained by the number of coursework hours. We also appoint an advisory committee for the student. It usually consists of the faculty advisor plus four other faculty members. At least one person from outside the department serves on the committee. The aim of this committee at the PhD level is to make certain that the students are solid biologists when they graduate. If an area of weakness is identified through examination of coursework taken as an undergraduate and a master's program, the student will be advised to take a course in that subject. All of my students must meet certain statistical requirements as well.

After 18–24 months in our department, PhD candidates are required to take what is called a preliminary exam, both written and orally. In my mind, this exam is the greatest hurdle in securing a PhD degree. The exam is given to "assess" the breadth and depth of basic biological information possessed by the student. The areas covered are usually those in which the advisory committee members are experts. A student who passes the written exam will then be subjected to an oral evaluation, which generally lasts about 3 hours. At least in part, the oral portion of the test is to not only examine with respect to breadth and depth of knowledge but also to determine if the student can think on their feet. When we began our PhD program in the late 1960s, financial support in terms of a teaching assistantship lasted 3 years, but about 20 years ago, it was extended to 4 years.

It is of interest that graduate programs in Europe are quite distinct from those here in the United States. Graduate students across the Atlantic take very few courses of any kind in pursuit of their PhD degrees (and very few of their colleges/universities offer master's degrees). It is assumed that their breadth and depth in terms of basic biological knowledge is adequately acquired as an undergraduate student. Most undergraduate programs in Europe require 3 years, with a 1-year honors program tacked on after the 3 years of coursework are completed. European undergraduate students majoring in biology take nothing but biology courses. If additional work in chemistry, physics, mathematics, or statistics is required, faculty in the biology department will teach the students. In addition to biology, chemistry, physics, and mathematics courses, our undergraduate institutions require a variety of liberal arts classes that may range from politics and economics to art and literature.

The European graduate programs also have no preliminary exams, or final dissertation defenses, like those administered in the United States. The student uses one external examiner for final approval of the dissertation. It was my privilege to be invited by Clive Kennedy to be the external examiner for John Aho when it was time for him to finish at the University of Exeter in England. After arrival in the city of Exeter, we spent time in the field discussing his collecting site and in the lab where he clarified some minor issues dealing with the content of his dissertation. That was it. I was especially excited about his field site because it was on the River Swincombe up in the very desolate hills of beautiful Dartmoor, not far from Exeter. After a careful inspection of his dissertation, I signed the document and returned home. John then left Exeter with his PhD and traveled to the University of Alberta in Canada where he began a postdoc with John Holmes.

Research

A master's student

Some beginning graduate students have had no research experience, which may present some real difficulties at first. Others acquired research experience as undergraduates and should be able to quickly adjust to life in graduate school. Whether a student has done any work as an undergraduate or not, I still consider all of them as "rookies." Accordingly, this is the time that master's students actually "learn" how to do research. For these students, I believe it is essential that the advisor watch them closely and carefully. Regular meetings of the advisor and student are a must. Periodic meetings of the student and advisory committee are also necessary. It is during this time that the student can acquire good habits or bad. Perhaps the most crucial requirement for success, however, is learning how to manage their time. There are many demands for the new students, and establishing priorities in the early part of graduate school is not always an easy thing to do.

Generally speaking, the master's student has slightly less than 2 years to take 24 hours of formal coursework, write a research prospectus, assemble equipment and supplies, become familiar with a pond or terrestrial habitat if fieldwork is a necessary component for their research, prepare a work description (if needed) for Animal Care and Use Committees, design experiments, analyze data, keep up with teaching assistantship duties (teach, grade papers, etc.), write a thesis, prepare and give a seminar, and defend the thesis. Throughout the MS process, the student will gain the experience they must have if they are going to take the next step in graduate school.

A frequent error made at this level occurs when the student "bites off more than they can chew." I know firsthand about these things, because I have come close to allowing a student to make this mistake. For example, elsewhere in the book, I described a difficult situation involving Brian Keas. He conducted some very intricate and complicated dietary studies on the growth and reproductive capabilities of uninfected *Helisoma (Hel.) anceps* and snails infected with *Halipegus (=Hal.) occidualis*. Unfortunately, I had no previous experience with this sort of research and encouraged him to do almost too much. Luckily, I stopped him from going too far, in time for him to obtain his master's degree within the allotted 24 months.

Another issue to consider arises when research does not produce the data hoped for during the beginning of a study. The advisor and student should be prepared for this potential adversity and have a secondary plan in place. Kyle Luth was faced with such a problem. However, in his case, we were not prepared with a backup strategy. He was fortunate we were able to devise a new research plan quickly, one that he was able to follow without "missing a step," and he completed his thesis work on time.

A PhD student

It is at this level that the student must blossom. As with master's students, proper use of time and identifying appropriate research objectives are absolutely necessary.

Mike Barger met with a unique problem when he began his dissertation research up on the Yadkin River headwaters. He quickly encountered a new opecoelid trematode species, which he felt obliged to describe before starting his dissertation fieldwork. To make his situation even more complicated, he also identified another opecoelid fluke (*Plagioporus sinitsini*), one that was progenetic in the snail *Elimia symmetrica*. Paedogenesis/progenesis is a fascinating concept/phenomenon and definitely worthy of a PhD dissertation. However, we were concerned about where the research would take him, how much hard data it would generate, and how long it would take. So, after some interesting, but brief, work with the fluke (he was able to write an appealing paper dealing with this research), he profitably pursued his dissertation research.

I have heard some folks say that the successful master's student will learn how to do research and that the PhD student should then exploit this skill. I wholeheartedly agree with this view. In my own experience, once I started at the master's level, my research effort moved smoothly, in part because I had done some research while an undergraduate. Some students have a tendency to wander off target and forget their objectives. I think another of the major tasks of an advisor at the PhD level is to encourage the student to acquire a set of "blinders" that will keep them focused on the mission at hand.

I also was lucky that I had a cadre of people to help in my early days as a PhD student. Dr. Self kept me on track, especially when I was exasperated enough to leave graduate school and move on to another career (my wife, Ann, also played a significant role in this regard—she is a very tenacious person). Cal Beames, a then well-advanced PhD student in the Department of Zoology, knew enough biochemistry to point me in the right direction when I needed to resolve a technical problem in achieving my goal, and MacWilson Warren helped me design and write a successful National Institutes of Health (NIH) predoctoral grant proposal. Not only did the grant take my wife and me out of virtual poverty, but also it helped by allowing me to focus on research without worrying about grading papers, monitoring exams, teaching in labs, etc.

All of my ecologically oriented PhD students learn a lot of statistics and even some modeling procedures, but not from me (I admit to being an "inadequate" statistician). However, I remember reading Harry Crofton's seminal papers when they appeared in 1971 in *Parasitology*. Frankly, because of my weak statistical/modeling background, I had a difficult time when I first read the papers. However, as I perused them again and again, I became satisfied that I knew enough about the phenomenon of overdispersion to understand the concept he had developed. The significance of Crofton's work was great enough in my mind that I began emphasizing the value of quantitative parasitology to my students from that point on. Students of mine, like Derek Zelmer, Nick Negovetich, Mike Zimmermann, and Dave Marcogliese, already had good math backgrounds from their undergraduate or master's-level experiences, and they were fine from the beginning. The others dug in and learned on their own. Fortunately, there were also faculty colleagues in our midst, for example, Dave Weaver in the Department of Anthropology, who willingly gave students much help and insight.

As mentioned previously, the financial support provided by a PhD teaching assistantship at Wake Forest lasted for 3 years early in our program but has since been extended to 4 years. The latter number is much better for the sort of work my students are typically engaged in since the data generated now spans three seasons in the field, a real bonus for the ecological types.

The PhD is the crown of graduate education. It is the highest graduate degree and takes a special kind of dedication and hard work to obtain it. As I have suggested previously, be certain regarding your choice of a mentor, the school/university you choose to attend, and the research you want to accomplish. I have also seen several students in our department change labs/mentors between their master's and PhD degrees. Do not be afraid to follow this course. Kym Jacobson switched from my lab into Ray Kuhn's; accordingly, her master's degree was oriented ecologically and her PhD was in immunoparasitology. Interestingly, she somehow hooked up with a federal (NOAA) marine lab in Oregon and has successfully changed back to ecological parasitology. Eric Wetzel conducted his master's research on a rhabditid nematode in flying squirrels in the lab of a very good ecologist (Pete Weigl) here at Wake Forest but moved to my lab to focus on *Halipegus eccentricus* that has a snail/microcrustacean/odonate/green frog life cycle. My point here is do not be afraid to make a change. Your career belongs to you!

The previous paragraph provides a nice segue into other problems frequently facing students. First, should an undergraduate student stay on at the same school to obtain a master's degree? Second, should a master's student stay at the same school to obtain a PhD degree? Third, should a student do an undergraduate degree and both graduate degrees at the same institution or even in the same lab? Let's look at the last question first because the answer is easy. No! We have faced this issue here at Wake Forest and resolved it. Students are not allowed to obtain all three degrees here, and certainly not in the same lab, nor with the same faculty mentor. If one of our undergraduates wishes to do their PhD at Wake Forest, then they must first go elsewhere to earn their master's degree. Then, they may come back to the same lab to work toward their PhD In a very few cases, students who have received an undergraduate degree elsewhere have come to Wake Forest to pursue a PhD degree directly and skip the master's degree. In some universities, students must follow this kind of program, but, personally, I think they are better off by acquiring a master's degree first. The experience of research and acquisition of the master's degree means that they can stop at this point if they choose or easily leave one institution and move on to another. I have also heard of institutions that will offer a master's degree if, after a couple years, a decision is made that they are not qualified to secure a PhD.

The first two questions above are also relatively easy to answer. Amy Crews was one of my undergraduate students who wanted to stay at Wake for master's degree and I encouraged her to do so, and then she went on to the University of Oklahoma where she worked toward her PhD with Tim Yoshino. When Tim moved on to the University of Wisconsin, Amy went with him. I actually encouraged Amy to stay here for both of her BS and MS degrees because I felt she would have a better chance for success at the PhD level if she had more research experience. This is why I also had her work with Mike Riggs while she was still a senior at Wake, that is, to gain experience. The plan worked well. She was the student who first discovered *Hal. occidualis* in Charlie's Pond and led the way for a "gaggle" of graduate students on a fascinating research expedition that lasted some 31 years!

Five of my students obtained both their master's and PhD degrees here with me. I also stayed with Dr. Self for both degrees at the University of Oklahoma. In my case, part of the reason was that my wife was pursuing her undergraduate degree while I was after both graduate degrees. We had our first child during my second year at Oklahoma, so she was tied down both with caring for a new baby and going to school. We felt she would lose course credits if I moved on to another school to do my PhD work. Moreover, and probably, the most significant factor in our decision was that we both felt comfortable as students there. I also thought that Dr. Self was an excellent mentor and we had already worked out a great dissertation project, one that ultimately helped me to obtain an NIH predoctoral fellowship.

The five guys who stayed with me for two graduate degrees I think felt the same as I when time came to make their decisions as to whether to move on or not after receiving their masters' degrees. I have not talked with them about the situation, except for Kyle Luth. While writing this segment of the book, I asked him about it. Why did he stay? In part, he remained at Wake Forest because we offered him, and the other four students, an interesting PhD research opportunity. Moreover, he was satisfied with both the prospective research work and the surroundings in which he was to live during the remainder of his stay.

These decisions ultimately rest with the student. There are several with whom I would have enjoyed working had they stayed on, but they had other plans. In these cases, I was always pleased to help them in whatever way I possibly could. All of them have had wonderful success in their careers, and I still claim them as my students!

The student's advisory committees

As mentioned briefly in previous text, advisory committees should play a significant role in securing a graduate education. Unfortunately, members of these committees are sometimes selected but then neglected. This is wrong. Yes, a mentor must be the primary person in keeping a student on track. However, the committee members also should be involved. They must be chosen in such a way that they play a role in advice regarding coursework taken but, more importantly, in the research planned.

However, be careful. Know the faculty before you pick them. Unfortunately, I have seen cases involving a couple of my faculty colleagues who have behaved in a very tactless manner for our students. In one example, a committee member refused to sign a dissertation until the student had made the precise changes dictated by the faculty member. It seemed as though the faculty member did not trust the graduate student to do as he requested. In another situation, just before a student's dissertation defense, I heard a faculty member refer to a foreign student as being "stupid" (the student passed). One needs to be aware there are people like this on most faculties. The best resolution for this kind of problem is to determine who these people are and not use them. The older students know about things like this—seek them out and learn who is reasonable and who is not!

Fortunately, the preponderance of faculty in most universities includes very decent men and women. Their interests are in our students. They want to see success and they will "go the distance" to assist these students in an ethical manner.

Special skill requirements

Graduate programs in most universities are similar. However, many have their own special requirements. When I arrived at Wake Forest, the highest degree offered in our department was the MS At that time, a special requirement for the student was to be familiar with a foreign language (most likely German or French) to be able to translate a scientific paper into English. When I was in graduate school, the special skill requirement for the PhD degree involved two languages. I was lucky enough, however, to persuade my advisory committee that there was probably a greater probability of encountering a scientific paper written in Spanish or French than in German.

When our department here began planning a PhD program, I was able to persuade my colleagues that language was no longer a useful special skill. Why? Because English had become the world's dominant scientific language, not German or French, as they had been during most of the 19th century. Thus, the language requirement disappeared in our department. However, it took us a while to convince some of the older faculty to abandon the old and worn-out language requirement, especially in the basic science departments of our medical school. They played a very negative role. For some reason, I have always felt there was an undeserved arrogance associated with some basic medical faculty members, especially the older ones, probably a residual feeling left over from my early experience when we were attempting to obtain a PhD program in our Department of Biology.

In place of the old language requirement, we introduced options for several new specializations. The ability to fix and embed and then section and stain tissues mounted on glass slides was found to be a special skill by some students. Other techniques, such as PCR and sequencing, electron microscopy, etc., would be useful for other students. Many have used advanced statistics, modeling, and other ecological methods for data analysis as special skills.

Handling the special exams

While putting together this special chapter, I began thinking about the preliminary exam. Earlier, I referred to it as perhaps the greatest hurdle in securing a PhD degree. So, I felt I should spend some extra time writing about it.

The first thing I would say is that after adequate preparation, most PhD candidates should be able to pass the exam without any difficulty. Perhaps the greatest problem for the student is in deciding what is meant by the "breadth and depth," or scope, required during preparation for the exam. The latter will depend on the kind of biological expertise possessed by each member of the examining committee. When I was in graduate school and faced my preliminary exam, any faculty member in the Department of Zoology could ask a question and even participate in the oral part of the exam if they desired. Think about this for a moment. It meant that hypothetically I had to be responsible for any subject in zoology, ranging from ecology to invertebrate systematics and from genetics to physiology. Fortunately, most of the Oklahoma faculty were reasonable and let my advisory committee to exclusively deal with my exam.

However, there were a couple of "snarky" old men who attempted to ambush me. I know this sounds somewhat paranoid on my part, but I admit that is how I thought about them. One of them was a physiologist who wanted to teach me a lesson for not taking one of his courses. What is the takeaway message here—it is kind of like I described earlier. You will encounter some bad guys along the way, so be prepared.

The second one was like the physiologist just described, only more perverse. He was a developmental biologist. I had taken an embryology course as an undergraduate, but Dr. Self decided that it would be a smart thing if I "sat in" on at least the lecture part of the embryology course. When I took the exam, I was asked just one question by the embryologist, "Describe the embryology of the eye." I was really bothered by this question for a couple of reasons. First, he had not yet given a lecture regarding the embryology of the eye, and I could not even begin to answer his question (I suspect he knew this would be the case). Second, he was not on my committee, and I was not prepared for a "shot" from him. My punishment for being ignorant about the embryology of the eye was that I had to take the rest of the course, sit in on the lab, and get A's on all the hour exams he gave during the rest of the semester. I was also not told what would happen if I did not do well (and I still cannot describe the embryology of the eye, and I do not recall if he even lectured on it during the remainder of the course). The "takeaway" here is do not sit in on courses unless you are receiving credit for them or unless you know the faculty member very well (especially if you are going to take the preliminary exam that same semester).

The real lesson learned from my two experiences is simple and has stayed with me throughout my career. First, the prelims should be fair. Second, only the advisory committee should administer the exam, and the members should provide at least a modicum of information regarding the focus of questions they will ask. It is always easy to flunk a student, but that should not be the objective of any exam! An examination can never reveal how much a student knows, only how much they do not know.

Writing and defending the thesis/dissertation

Unfortunately, writing and public speaking are learned skills—you are not born with either one. However, it seems to me that a good number of people have a knack for expressing themselves very well—some can do it verbally, some in writing, and others are lucky because they are able to do both. For the last couple of years, I have taught a first year seminar (FYS) titled "The creation of Darwin's theory." Mine is but one of the many FYS from which our freshmen students can select during their first year. These are not "content" courses, but are focused on writing and, to a lesser extent, on speaking. Over these 2 years, I have found that about 25% of our incoming freshmen are pretty good writers and some of them possess the ability to speak very coherently. A few are able to express themselves well using both venues. I cannot say for sure that the rest of them will become reasonably good writers or speakers, but my guess is that most of them will, if they are required to do a lot of both.

When it comes to writing a scientific paper, a thesis, or a dissertation, I learned early that there was a correct, and an incorrect, way to do it. As an undergraduate, I was required to compose several relatively short pieces, or essays, but nothing like the very first real scientific paper I wrote, which was based on the results of my dissertation. When I received the reviews from that first attempt, I was rather dismayed/distressed. At the time, I was doing my postdoc with Dr. John Larsh at UNC–Chapel Hill. Since he was a prolific author, Dr. Self was a thousand miles away, and "snail mail" was relatively slow, especially in those days, I approached Larsh and asked if he would take a look at the reviews and help me repair the damage. A few days later, he invited me to come to his office, and we discussed what I had done incorrectly. His explanation was really quite simple but revealing. Very simply put, I had written the paper using the wrong course of action.

He explained that the correct way to write a scientific paper is to first organize the figures and tables in the sequence I wanted them to appear in the text, assuming of course that I had analyzed the data correctly so that the reader could easily interpret them. As he explained, once this is done, the writing should be easy. Just as important, however, the Results section should be written first, followed in sequence by the Discussion, Introduction, and Materials and Methods, with the Lit Cited section and Abstract completing the paper. The problem with my first attempt was that

I had written the Introduction first, followed by the Materials and Methods, Results, and Discussion, in that order—the correct sequence for a published paper but the wrong one for preparing a thesis, dissertation, or paper for publication! My data analysis and figures were fine. However, in my first Introduction, I had inserted a couple of objectives/goals that were not part of the Results. When I wrote the paper a second time, beginning with the Results section, it made complete sense to focus on the data I had actually generated, not those that I wanted to see. If you write a paper the way Dr. Larsh suggested, the target will be hit every time.

I do not pretend to be a brilliant writer, but that is okay because all I really want to do is be a good writer. I have never been taught by anyone about the technicalities of writing, for example, when to write, how long I should sit each time I decide to write, etc. However, I believe I have done enough writing and editing in my career to offer some suggestions that have assisted me over the years. First, do not be afraid to start a paper, but only if you have sufficient data to write about or a good story to tell. Some people are literally too fearful to write. Do not begin a paper feeling this way—I promise that the more you write, the better you will become at writing. Second, I have heard it said that the secret to good writing is rewriting. I now strongly believe in this old adage. When you rewrite something, you are editing your own work. How many times should you edit your paper? I would say until your writing sounds like you. This may sound somewhat nonsensical, but it is true. The longer you write during a career, the easier it will become. Try reading it out loud. When it sounds like you, stop editing. Third, find someone who is willing to objectively critique your work. Note that I said critique, not criticize; these two words are antonyms, not synonyms. The former means to assess or evaluate; the latter means to condemn or disparage. The purpose of a critique is to construct, not tear down.

I was editor of the *Journal of Parasitology* for 19 years. One of the things I learned is that even the best people in my field of parasitology do not write perfect papers. Most authors accepted my critiques, but a few others were quite unwilling. Eventually, I was able to convince even the best of our writers/parasitologists that my revision might be better than what they had written. When you obtain a critique, accept it as a body of suggestions. You do not need to agree with all of their ideas, but you should at least consider all of them. Finally, once you think you have it the right way and before you send the thesis/dissertation to your advisor, or your committee of advisors, or the manuscript to a journal, let it "cool off" in your top desk drawer before you part with it—but always then read and edit it one more time before you send it (just like I am doing now).

I think one of the best approaches to writing is to first get your ideas down on paper or into your computer. If you can write three to five pages a day, you are making good headway, which is what you obviously want. The next day after creating the first three to five pages, start your work by editing the three to five pages you wrote yesterday. When the editing is complete, then write another three to five pages. The next day, repeat the editing from the beginning, before you start writing fresh material. As you write new material and edit the old, you will begin to see a better and better flow of information. The story you are telling will come together more clearly and make it easier to follow.

I think my writing skills have improved over my 50-year career. I actually enjoy doing it now!

Publishing papers and selection of the journal

If research is worth doing, then so is publishing it. Do not let it sit. Why? Because someone may trump your research, and where does that leave you—sitting alone with a lot of wasted effort down the drain. In addition to conveying the results of your research, you also want to establish priority for your work. I should note that both your thesis and dissertation are considered as publications, which is why they are frequently cited in recognized scientific journals. In part, this is because they have both been subjected to peer review, that is, your advisor and advisory committee have certified their approval by signing the document.

The first decision you make after obtaining sufficient new data is where will it be published. Where you publish is your choice, although your advisor may definitely have a say in the option of journals. Listen to what this person tells you. Even though you may still be in the process of collecting data for your master's degree or PhD, you are a rookie. Another determination you make must be, will there be coauthors? My feeling about this issue is similar to that of many other academics. If your advisor makes a genuine intellectual contribution to the thesis or dissertation, then they should be included as a coauthor, not the senior author. Another of my policies over the years has been that if the student does not publish a thesis or dissertation within 5 years of graduation, then I will write the paper with the student as senior author (again, if the research is worth doing, then so is publishing it). On occasion, other students may appear on the publication. I think this is absolutely okay too, so long as the other student contributed directly to the study. This has not happened in all cases with my students, but it has for a few.

The selection of a journal for publication is a very important step. A number of academics, especially in the sciences, make their journal decision based on something called an impact factor (IF). An IF is a metric used for calculating the "supposed" significance, or "quality," of a given journal. The IF for a journal in 2014 would include the number of times that all papers published in the journal in 2012 and 2013 are cited by indexed publications during 2014 divided by the total number of citable papers published by that journal in 2012 and 2013. This procedure makes the denominator of the equation very meaningful in calculating the IF. Journals that publish research notes are a disadvantaged because these very short papers increase the size of the denominator while simultaneously reducing the size of the numerator. Why? Because research notes are less likely to be cited than a regular article. In contrast, journals that publish review articles with a high degree of frequency are much more likely to have higher IFs because a review is much more likely to be cited.

Any author, whether a rookie or an established scientist, would want to see their papers published in journals with relatively high IFs. Accordingly, some people select their journal for publication based on the IF. I have even heard it said that the strength of a journal IF is so significant that they may actually affect faculty raises, promotion decisions, and success with granting agencies in some universities and countries, especially in Europe.

Having been an editor for so long has caused me to develop a great personal contempt for selecting a journal if the IF is the primary criterion for quality assessment. The data should always match the journal. My feeling is that the choice of a journal in which one should publish should be based in large part on the primary scientific society to which the author belongs. Since I am a member of the American Society of Parasitologists, the *Journal of Parasitology* has been my primary choice for publication during the past 25–30 years. I have, however, also continued to publish papers in *Parasitology, International Journal for Parasitology, Acta Parasitologica, Journal of Tropical Medicine, Journal of Veterinary Medicine, Systematic Parasitology,* and *Comparative Parasitology,* among others. My final decision regarding journal choice is also based on special content of the paper, readership of the journal, the turnaround time of the journal, etc.

When a PhD student conducts research in my field of parasitology, it is frequently accomplished in "bunches." If the work is done in this manner, I generally encourage the student to publish as they proceed with their research. It is always a good thing if you have a "string" of publications when you finish your degree. It should also be emphasized that PhD dissertations are unique individually and that some liberties in presentation style can be made that would not be acceptable in writing for a journal publication. For example, redundancy in a dissertation is acceptable, but not so for a journal publication. I do not think it is a good idea to wait until the dissertation work is completed before writing of a journal paper begins. I also believe it is completely legitimate to break the dissertation up into functionally related papers and not to write one very large paper. Some folks might object to this approach and claim that a group of smaller papers rather than one or two large ones is simply a way of artificially expanding the size of one's curriculum vitae (CV). This method could be considered as padding a CV, but it need not be if the papers are appropriately written.

Another point should be made here. Once you have decided on a journal, you must carefully read its publication guidelines, and I mean to really consider them carefully! This is exceedingly important. The editor, associate editor, and referees are all busy people. Sloppy writing, or failure to follow the guidelines, for example, even regarding the placement of a comma, or bold font for a colon or a subheading in the text, is a frustrating error for those who are responsible in generating a set of papers ready to send away to the copy editors and the publisher of a journal. A large number of typos, misspellings, and other technical mistakes frequently may even make those in the process of reviewing or making decisions regarding publication of the paper wonder about the accuracy of the data. I realize these sorts of errors in a manuscript seem like they should be considered as very minor. They are when they are few in number, but some authors just cannot, or will not, understand the significance of the time it takes to edit a paper and end up with a "sloppy" paper.

So, do not allow yourself to make these sorts of "mystakes!" Be "kareful!"

Collaboration

Collaboration is an important way of doing science, especially in today's world of specialization. The advantage of collaboration is clear, especially when unique technology, or instrumentation, or skills are required to achieve research success. An important rule for collaboration is to be sure that only "hands-on," or intellectual, contributors are included as coauthors on a publication. The senior author, who is generally the leading investigator on the project, must make the decision regarding inclusion. The senior author also decides the sequence of collaborators.

Graduation (and sometimes a postdoc experience)

As I have stated earlier, the crown of graduate work is a PhD. If you expect to teach and do research in a major university anywhere in the world, then postdoc experience is a necessity. Funding is not always easy to acquire, making the postdoc requirement a very competitive and difficult route to follow. Many of the smaller colleges and universities will require "experience," without stipulating that it is as a postdoc. Some universities, like Wake Forest, are now hiring PhD's with postdoctoral experience into tenure-track positions, along with PhD's without postdocs into nontenure-track jobs that are partly teaching and partly administrative. I am not sure I like this way of doing things, but the administrations seem to like it as a way of reducing costs and creating better opportunities for tenure-track faculty to have more time to conduct research while simultaneously teaching reduced loads.

Would you do it again?

I cannot answer this question for you. However, for me, this is an easy question to answer. Yes, I would do it again and again and again. In the first place, after securing my graduate degrees and more than 50 years in the academic world, there is no way that I can visualize myself in another occupation. In another chapter, I told the story about asking Doc Stabler why he chose an academic career and he responded,

"Three reasons, [pause] June, July, and August." If wisely used, the opportunities presented by these 3 months every year are immeasurable. Personally, an academic career has also given me the chance of interacting with an inordinate number of interesting and exciting students and faculty colleagues.

Joseph Joubert, an 18th-century French writer, said, "To teach is to learn twice." Teaching, when coupled with the inquiry of research, should represent the nearly complete fulfillment of any academic career. The only thing missing from such a statement is that some of us have been able to fully enrich our scientific lives by interacting with a cadre of wonderful graduate students and interesting professional colleagues throughout our careers.